Introduction to Digital Control of Linear Time Invariant Systems

Introduction to Digital Control of Linear Time Invariant Systems

By

Ayachi Errachdi

Cambridge Scholars Publishing

Introduction to Digital Control of Linear Time Invariant Systems

By Ayachi Errachdi

This book first published 2022

Cambridge Scholars Publishing

Lady Stephenson Library, Newcastle upon Tyne, NE6 2PA, UK

British Library Cataloguing in Publication Data
A catalogue record for this book is available from the British Library

ISBN (10): 1-5275-8271-X
ISBN (13): 978-1-5275-8271-2

TABLE OF CONTENTS

Preface ix

About the Author xi

Chapter One 1
Sampling and reconstruction
1.1. Introduction
1.2 Digitization of continuous-time signal
1.2.1 The sampling of continuous-time signal
1.2.1.1 Ideal sampler
1.2.1.2 Practical sampler
1.2.1.3 Commonly used functions
1.2.1.4 Sampling period choice
1.2.2 The quantization of sampled signal
1.2.2.1 Uniform quantization
1.2.2.2 Non uniform quantization
1.2.2.3 Signal to quantization noise ratio
1.2.3 The coding of quantized signal
1.3 Reconstruction of analog signal
1.3.1 Zero order hold
1.3.2 First order hold
1.3.3 Cubic order hold interpolation
1.4 Applications
1.5 Conclusions
1.6 References

Chapter Two 48
The z-transform of systems
2.1 Introduction
2.2 The z-transform
2.2.1 Relation between Laplace transform and z-transform
2.2.2 The region of convergence
2.3 The z-transform by residual method
2.4 Commonly used z-transform functions
2.4.1 Unit impulse function

2.4.2 Unit step function
2.4.3 Unit ramp function
2.5 Properties of the z-transform
2.6 Examples of z-transform
2.7 The inverse z-transform
2.7.1 The inverse z-transform by the inspection method
2.7.2 The inverse z-transform by the direct division method
2.7.3 The inverse z-transform by the partial-fraction expansion method
2.7.4 The inverse z-transform by recurrence inversion method
2.7.5 The inverse z-transform by contour integration method
2.8 Examples of inverse z-transform
2.9 Conclusions
2.10 References

Chapter Three.. 79
The discrete-time transfer function
3.1 Introduction
3.2 Transfer function from impulse response
3.2.1 Impulse function
3.2.2 Impulse response of system
3.2.3 Calculating frequency response
3.3 Transfer function from difference equation
3.3.1 Difference equation models
3.3.2 The z-transfer function
3.3.3 Difference equation representation
3.3.4 From z-transfer functions to difference equation
3.4 Rational z-transforms
3.4.1 Poles and zeros
3.4.2 Pole and zero location and time-domain behavior
3.5 Transfer function of continuous system follows a zero-order hold
3.6 Transfer function of block diagrams
3.6.1. Rules for block diagram manipulation
3.6.2. Influence of the position's sampler
3.6.3. Multiplied coefficient of discrete-time systems
3.6.4. Addition of discrete-time systems
3.6.5. Cascaded separated discrete-time systems
3.6.6. Cascaded continuous-time systems
3.7 Transfer function of closed-loop discrete-time systems
3.7.1. Two systems not separated by sampler
3.7.2. Two systems separated by sampler

3.7.3. Three systems not separated by sampler
3.7.4. Three systems separated by sampler
3.7.5. The sampler before the second system
3.8 State space description and z-transfer function
3.8.1. Definition
3.8.2. State space representation of a 1st order discrete system
3.8.3. State space representation of a 2nd order discrete system
3.8.4. State space representation of a nth order discrete system
3.9 Examples of transfer function
3.10 Conclusions
3.11 References

Chapter Four **120**
Stability of discrete-time system
4.1 Introduction
4.2 Stability definition
4.2.1 Stability of the 1st order system
4.2.2 Stability of the 2nd order system
4.3 Algebraic stability criterion
4.3.1 The Jury stability criterion
4.3.2 The Routh stability criterion
4.4 Domain frequency stability
4.4.1 Bode diagrams for continuous-time systems
4.4.2. Nyquist stability theory in continuous-time systems
4.4.3. Gain and phase margins in the Nyquist plot in continuous-time systems
4.4.4. Bode diagrams for discrete-time systems
4.4.5. Discrete Nyquist stability criterion
4.5 Root locus method
4.5.1. Position of poles and zeros of system in s-domain in open loop
4.5.2. Position of poles and zeros of system in s-domain in closed loop
4.5.3. Property and construction rules in s-domain
4.5.4. Position of poles and zeros of system in z-domain in closed loop
4.5.5. Property and construction rules in z-domain
4.6 Steady state error of discrete-time system
4.6.1 Definition of steady state error
4.6.2 The static position error constant
4.6.3 The static velocity error constant

4.6.4 The static acceleration error constant
4.7 Examples of applications
4.8 Conclusions
4.9 References

Chapter Five 175
Synthesis of digital controller for linear time invariant system
5.1 Introduction
5.2 Digital controllers
5.2.1 Proportional action
5.2.2 Integral action
5.2.3 Derivative action
5.2.4 Filtering the derivative action
5.2.5 Proportional-integral action
5.2.6 Proportional-derivative action
5.3 Proportional Integral Derivative controller
5.4 Tuning methods of controller
5.4.1 Dominant pole
5.4.2 Pole placement method
5.5 Application of digital PID controller for DC Motor Speed
5.6 Examples of applications
5.7 Conclusions
5.8 References

PREFACE

New technologies in engineering, automatics and robotics are creating problems in which control plays a major role. Solutions to many of these problems require the use of digital signals. This manuscript attempts to provide the reader with an insight into digital control of time-invariant linear systems.

My objective is to offer an accessible, self-contained research monograph which can also be used as a graduate text. The material presented, in this book, is of interest to a wide population of students, teachers, engineers, and researchers working in engineering, computing, electronic, robotics and automatics. It can also be used as a reference book by control engineers in industry and research students in automation and control.

The first chapter covers fundamental concepts in the sampling and reconstruction of signals. The material presented in the second chapter can serve as an advanced text for courses on z-transform and inverse z-transform. Indeed, the inspection method, the direct division method, the partial-fraction expansion method, the recurrence inversion method and the contour integration method are all detailed. The third chapter introduces the transfer function. In fact, the absence or presence of an input sampler is crucial in determining the transfer function of a system. For this reason, different examples of the position of the sampler are treated to improve its efficiency and its influence. The fourth chapter presents the stability condition of discrete-time systems in the closed loop. The global stability definition, the algebraic stability criterion and the stability in the frequency domain are discussed. The fifth chapter introduces the synthesis of a digital controller for linear time invariant system. The last section, in this book, shows the use of a digital PID controller in the practical speed control of a DC motor using an Arduino card, to encourage readers to explore new applied areas of digital control. In all these chapters simple examples are used to illustrate important concepts.

I hope that the publication of this work will have a positive impact on students' interest in the subject. I have been benefited from my students, through my teaching and other interactions with them; in particular, their questions asking me to explain many of the topics covered in this book with simple examples.

Ayachi ERRACHDI, University of Kairouan, Kairouan, 2021

ABOUT THE AUTHOR

Ayachi ERRACHDI studied at the National Engineering School of Monastir, Tunisia, obtaining a degree in electrical engineering and a master's degree in automatic and industrial maintenance in 2005 and 2007, respectively. He obtained his PhD degree in electrical engineering from the National Engineering School of Tunis El Manar, Tunisia, in 2012. He is currently an associate professor of electrical engineering at the University of Kairouan.

He is currently a member of the Automatic Research Laboratory, at the National Engineering School of Tunis El Manar, Tunisia. His major research interests are in nonlinear, continuous and discrete-time control systems, artificial neural networks, system identification, adaptive control systems, fractional order systems, observers, simulation and target tracking. He has published many papers in journals and conference proceedings, and has supervised many doctoral students.

CHAPTER ONE

SAMPLING AND RECONSTRUCTION

1.1. Introduction

In this chapter we are going to focus on the sampling and reconstruction of an analog signal. Indeed, a continuous-time signal is an infinite and uncountable set of numbers. Between a start and end time, there are infinite possible values for the time t and the instantaneous amplitude. When continuous-time signals are brought into a computer, they must be digitized. In a discrete-time signal, the number of elements in the set, as well as the possible values of each element, is finite, countable, representable by computer bits and can be stored on a digital storage medium.

Digital systems attempt to overcome the analog system's susceptibility to noise by sacrificing the infinite aspect of the time and the amplitude resolution to obtain perfect reproduction of the signal no matter how long it has been stored or how many times it has been duplicated.

The discrete time and discrete amplitude nature of the digital signal provide a buffer to any noise that may enter the system through transmission or otherwise. Digital signals are usually stored and transmitted in the form of ones and zeros. If a digital receiver knows that only zeros or ones are being transmitted and when approximately to expect them, there is a certain acceptable level of noise that the receiver can handle.

Beyond the advantages of noise robustness during reproduction and transmission, digital signals have many other advantages. These include the ability to use computer algorithms to filter the signal, data compression to save storage space and signal processing to extract information that may not be possible through manual human analysis. Thus, there can be a large benefit in converting many signals that are used in cardiology to digital form.

1.2. Digitization of continuous-time signal

An analog signal is a continuous function with respect to the time and amplitude variables.

To find a digital signal, three steps are needed: sampling, quantization and coding as represented by the following figure.

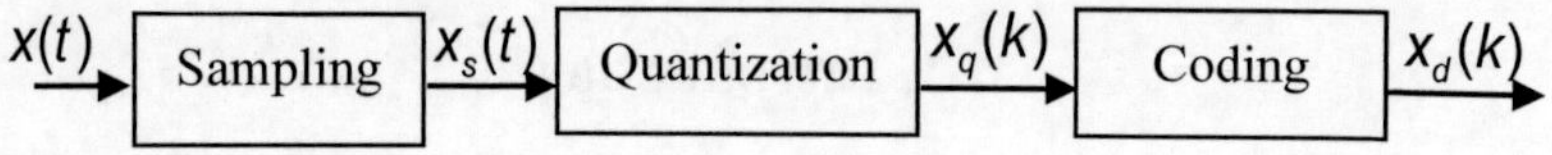

Fig. 1.1 Three processes of digital processing

where $x(t)$ is an analog signal, $x_s(t)$ is a sampled signal, $x_q(k)$ is a quantized signal and $x_d(k)$ is a digital signal.

These blocks are defined as follows:

- sampling (sampler block): using a sampler block, we find a sampled signal that is discrete in time and continuous in amplitude.
- quantization (quantizer block): using a quantizer block, we find a quantized signal that is discrete in time and discrete in amplitude.
- coding (encoder block): each sample quantized to a finite number of bits.

1.2.1. The sampling of continuous-time signal

The sampling process converts a continuous-time signal to a discrete-time signal with a defined time resolution. This is determined by what is known as the sampling rate, and it is usually expressed in Hertz (Hz) or samples per second. The sampling rate needed for a faithful reproduction of the signal depends on the fluctuation sharpness of the signal that is being sampled.

1.2.1.1. Ideal sampler

To sample an analog signal, means to register some values of this signal at given times. An ideal sampler is generally represented by an interrupter. The time closing is considered equal to zero. The ideal sampler is given by Figure 1.2:

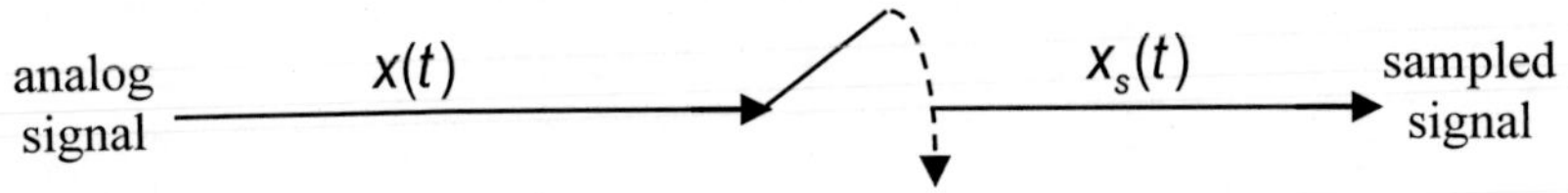

Fig. 1.2. The ideal sampler of an analog signal

We obtain a sampled signal $x_s(t)$ at equally spaced times $t = kT_s$:

$$x_s(t) = x(kT_s) \qquad \forall k \in]-\infty, +\infty[$$

where T_s is called the sampling period and it is inversely related to the sampling rate F_s, that is

$$F_s = \frac{1}{T_s}$$

The sampled signal $x_s(t)$ is found by multiplying the continuous-time signal $x(t)$ by a series of unit impulses, which are called the Dirac comb, given by

$$x_s(t) = x(t)p(t)$$

where $p(t)$ is

$$p(t) = \sum_{k=-\infty}^{+\infty} \delta(t - kT_s)$$

and $\delta(t)$ is called the Dirac delta function and is defined by:

$$\delta(t) = \begin{cases} 1 & if \quad t = 0 \\ 0 & if \quad t \neq 0 \end{cases}$$

The Dirac delta function $\delta(t)$ is a unit pulse in which the duration approaches zero but the area of the pulse is equal to one. This means as the width of the pulse τ approaches to zero, the amplitude of the pulse $\frac{1}{\tau}$ must approach infinity to maintain a unity area. The Dirac delta function is presented by this figure:

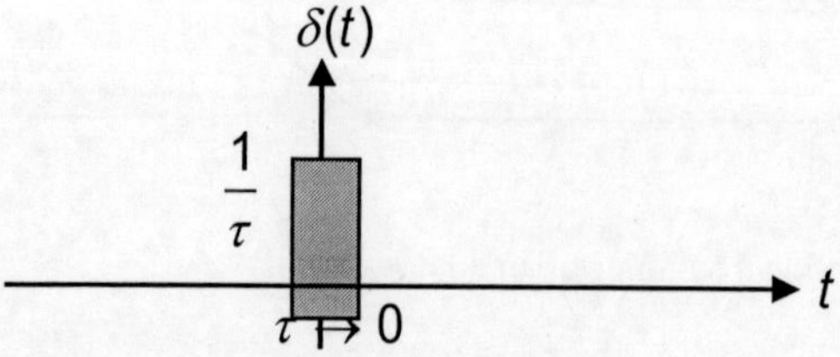

Fig. 1.3. The Dirac delta function

The periodic train impulse $p(t)$ is presented as follows

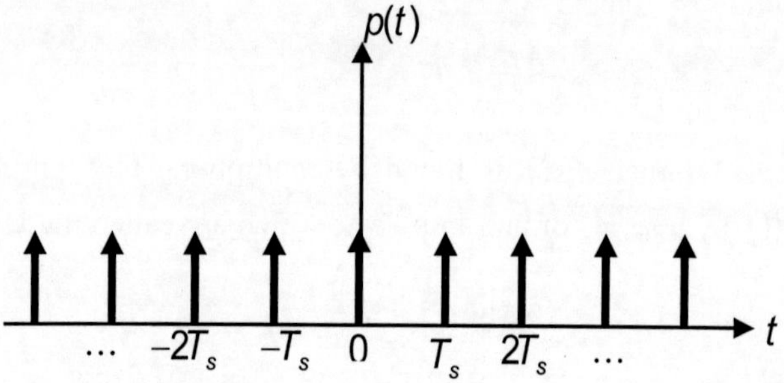

Fig. 1.4. Periodic impulse train

Sampling $x(t)$ is equivalent to multiplying it by a train of impulses as given here:

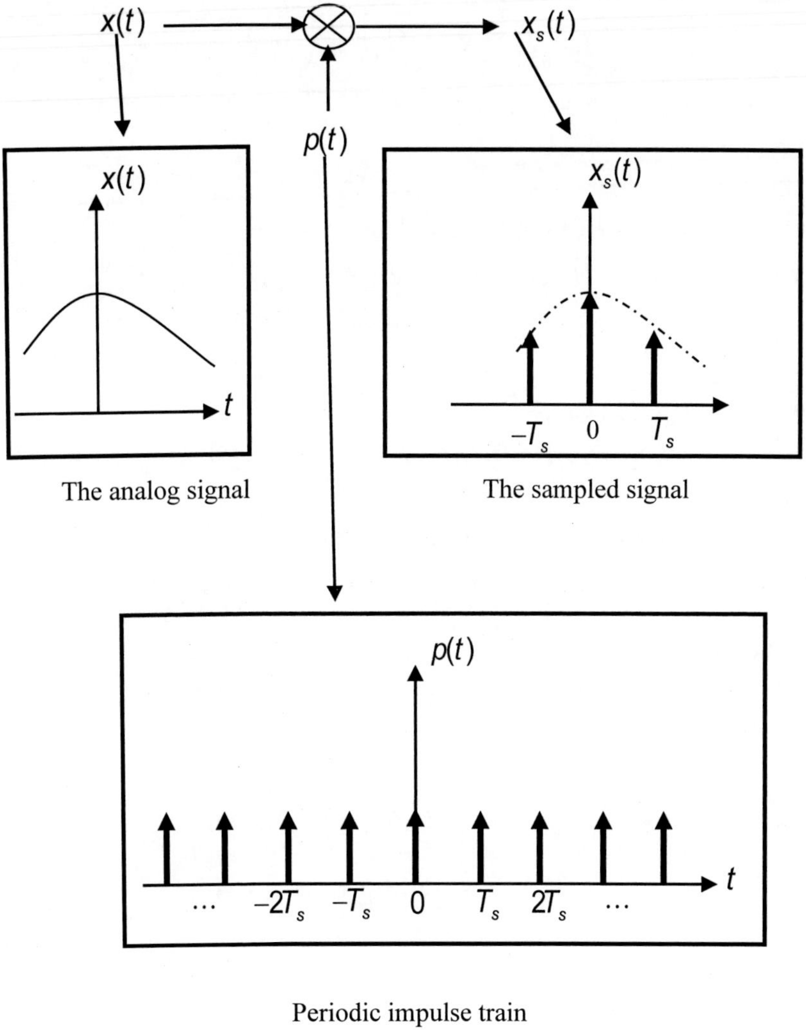

Fig. 1.5. The sampling of an analog signal

The found sampled signal is given as:

$$x_s(t) = \sum_{k=-\infty}^{+\infty} x(kT_s)\delta(t - kT_s)$$

whereas the continuous-time signal $x(t)$ is supposedly causal, that is to say

$$x(t) = 0 \qquad \forall \ t < 0$$

Therefore, the sampled signal is given as:

$$x_s(t) = x(0)\delta(t) + x(T_s)\delta(t - T_s) + x(2T_s)\delta(t - 2T_s)+, \ldots, +x(kT_s)\delta(t - kT_s)$$

$$x_s(t) = \sum_{k=0}^{+\infty} x(kT_s)\delta(t - kT_s)$$

Some properties are described as follows:

$$\int_{-\infty}^{+\infty} \delta(t - kT_s)dt = 1 \quad , \qquad \forall k \in \mathbf{R}$$

The Laplace transform of the Dirac function is

$$L\left[\delta(t - kT_s)\right] = e^{-kT_s s} \quad , \qquad \forall k \in \mathbf{R}$$

and the Laplace transform of the sampled signal is

$$L\left[x_s(t)\right] = L\left[\sum_{k=0}^{+\infty} x(kT_s)\delta(t - kT_s)\right]$$

$$= \sum_{k=0}^{+\infty} L\left[x(kT_s)\delta(t - kT_s)\right]$$

$$= \sum_{k=0}^{+\infty} x(kT_s)L\left[\delta(t - kT_s)\right]$$

$$= \sum_{k=0}^{+\infty} x(kT_s)e^{-kT_s s}$$

1.2.1.2. Practical sampler

In the practical case $\forall k \in \mathbf{R}$, the interrupter has an important closing time which leads to samples depending on the duration of the sampling λ.

$$x_s(t) = \sum_{k=0}^{+\infty} x(kT_s)\lambda\delta(t - kT_s)$$

where λ is the duration of sampling caused by the interrupter.

1.2.1.3. Commonly used functions

a. Unit impulse

Consider a continuous-time impulse $\delta(t)$, defined by:

$$\delta(t) = \begin{cases} 1 & if \quad t = 0 \\ 0 & if \quad t \neq 0 \end{cases}$$

The sampled impulse is:

$$\delta_s(t) = \sum_{k=0}^{+\infty} \delta(kT_s)\delta(t - kT_s) = \begin{cases} 1 & if \quad k = 0 \\ 0 & if \quad k \neq 0 \end{cases}$$

and is given by the following figure:

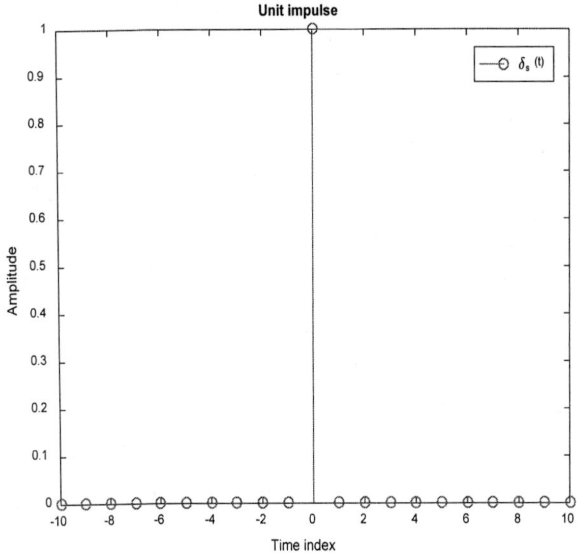

Fig. 1.6. The unit impulse

b. Unit step

Consider a continuous-time signal $u(t)$, defined by:

$$u(t) = \begin{cases} 1 & if \quad t \geq 0 \\ 0 & if \quad t < 0 \end{cases}$$

The sampled unitary step is:

$$u_s(t) = \sum_{k=0}^{+\infty} u(kT_s)\delta(t - kT_s) = \begin{cases} 1 & if \quad k \geq 0 \\ 0 & if \quad k < 0 \end{cases}$$

and is represented by the following figure:

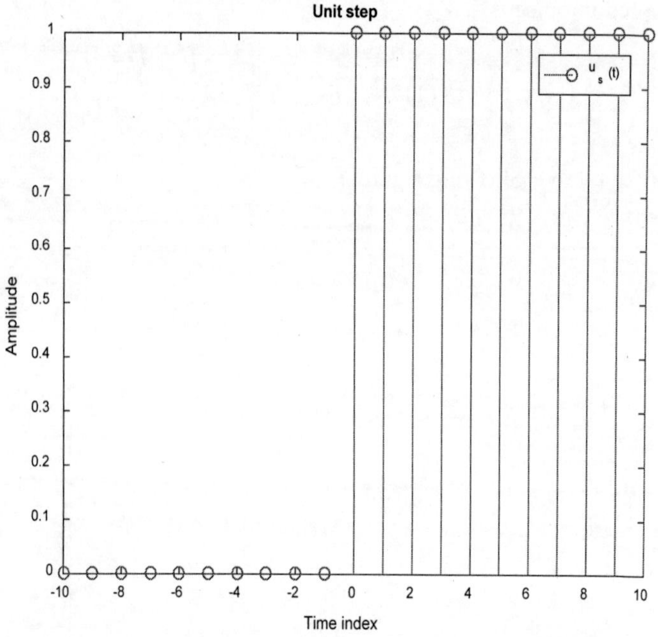

Fig. 1.7. The sampled unitary step

c. Unit ramp

Consider a continuous-time signal $r(t)$, defined by:

$$r(t) = \begin{cases} t & if \quad t \geq 0 \\ 0 & if \quad t < 0 \end{cases}$$

The sampled ramp is:

$$r_s(t) = \sum_{k=0}^{+\infty} r(kT_s)\delta(t - kT_s) = \begin{cases} k & if \quad k \geq 0 \\ 0 & if \quad k < 0 \end{cases}$$

and it is given by the following figure:

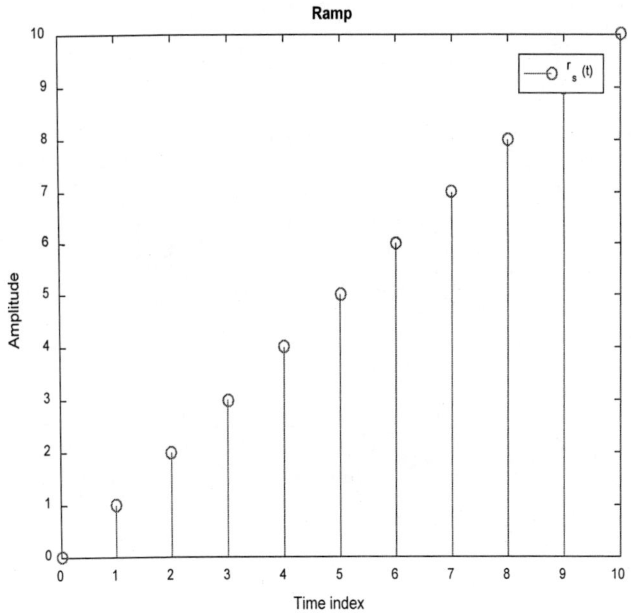

Fig. 1.8. The sampled ramp

d. Sinusoidal signal

Consider an analog signal $s(t)$, defined by:

$$s(t) = A\sin(wt + \varphi)$$

where A is an amplitude, w is an angular frequency (radians/s) and φ is a phase (radians). The expression of the sampled sinusoidal signal is:

$$s_s(t) = \sum_{k=0}^{+\infty} s(kT_s)\delta(t - kT_s) = A\sin(wkT_s + \varphi)$$

and it is given by the following figure:

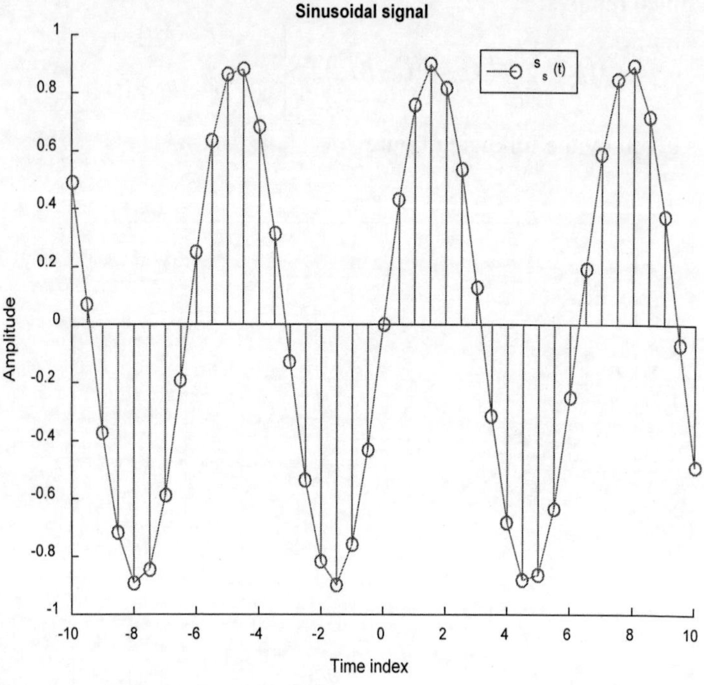

Fig. 1.9. The sampled sinusoidal signal

1.2.1.4. Sampling period's choice

A small amount of information is lost by sampling an analog signal at a very short time but all the information contained in the signal can be lost when the sampling times are too spaced from each other.

Example: Consider the analog signal defined by the following function:

$$x(t) = \sin(2\pi ft)$$

This function is sampled at a frequency $F_s = 2f$, that is, at a sampling period $T_s = \frac{T}{2}$.

$$x_s(t) = \sum_{k=0}^{+\infty} x(kT_s)\delta(t - kT_s) = \sum_{k=0}^{+\infty} \sin(\pi k)\delta(t - kT_s) = 0$$

An analog signal $x(t)$ having a low-pass type spectrum of width w_{max} is fully described by the further completeness of its instantaneous values $x_s(t)$ if they are elevated to a sampling pulse w_s such that $w_s > 2.w_{max}$.

a. Theorem of spectrum concept

The Fourier transform of a bandlimited sampled signal $x_s(t)$ is the frequency spectrum of the signal:

$$X_s(f) = F(x_s(t))$$

$$= \frac{1}{T_s} \sum_{k=-\infty}^{+\infty} X(f - kF_s)$$

$$= F_s \sum_{k=-\infty}^{+\infty} X(f - kF_s)$$

Proof:

The impulse function is given by the following equation

$$p(t) = \sum_{k=-\infty}^{+\infty} \delta(t - kT_s)$$

The impulse function has a periodic distribution of period T_s from which it can be decomposed into a Fourier series:

$$p(t) = \sum_{k=-\infty}^{+\infty} \delta(t - kT_s) = \sum_{n=-\infty}^{+\infty} c_n e^{\frac{jn\pi}{T_s}t}$$

with

$$c_n = \frac{1}{T_s} \int_{-\frac{T_s}{2}}^{\frac{T_s}{2}} p(t) e^{\frac{jn\pi}{T_s}t} dt$$

Using the property of the Dirac distribution:

$$\int_a^b f(t)\delta(t-kT_s) = \begin{cases} f(kT_s) & if \quad a < kT_s < b \\ 0 & if \quad no \end{cases}$$

it can be easily shown that $C_n = \frac{1}{T_s}$, from where, we deduce that:

$$p(t) = \frac{1}{T_s}\sum_{n=-\infty}^{+\infty} e^{\frac{jn\pi}{T_s}t}$$

Let us apply the Fourier transform to the expression of the sampled signal $x_s(t)$. This would be transformed to:

$$X_s[f] = F[x_s(t)] = F[x(t)p(t)]$$

Using the property of the impulse function and the linearity of the Fourier transform, we obtain:

$$X_s[f] = F[x_s(t)] = \frac{1}{T_s}F\left[x(t)\sum_{k=-\infty}^{+\infty} e^{\frac{jk\pi}{T_s}t}\right] = \frac{1}{T_s}\sum_{k=-\infty}^{+\infty} F\left[x(t)e^{\frac{jk\pi}{T_s}t}\right]$$

By the use of the spectral translation property of the Fourier transform, a shift in the spectral domain corresponds to a multiplication by a complex exponential in the time domain, according to the formula:

$$F\left[e^{j\pi F_s t}x(t)\right] = X(f-F_s)$$

The Fourier transform of the sampled signal becomes

$$X_s(f) = F(x_s(t)) = \frac{1}{T_s}\sum_{k=-\infty}^{\infty} X(f-kF_s)$$

We note that $X_s(f)$ is periodic with period T_s and it is obtained by the sum of an infinity of complex functions, each of them being the Fourier transform $X(f)$ of the continuous-time signal $x(t)$, shifted by kF_s where F_s is the frequency of sampling and $k \in \mathbf{Z}$.

The following figures illustrate the frequency-domain representation of sampling in the time domain. The spectrum of the original signal is presented in Fig. 1.10. Three cases are then to be considered according to the value taken by the sampling frequency compared to the highest frequency F_{max} contained in the spectrum $X(f)$ of $x(t)$. The Fourier transform of the sampled signal with $F_s > 2F_{max}$ is shown in Fig. 1.11. The Fourier transform of the sampled signal with $F_s = 2F_{max}$ is shown in Fig. 1.12. The graphical representation using the Fourier transform of the sampled signal where $F_s < 2F_{max}$ is given by Fig. 1.13.

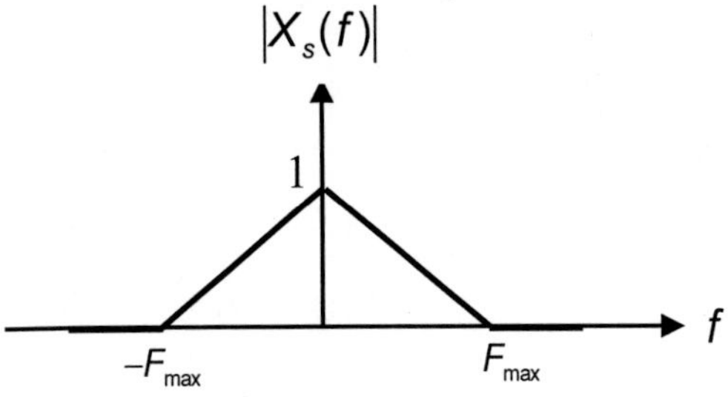

Fig.1.10. The spectrum of the original signal

Case 1: If $F_s > 2F_{max}$, the Fourier transform of the sampled signal is shown in Fig. 1.11.

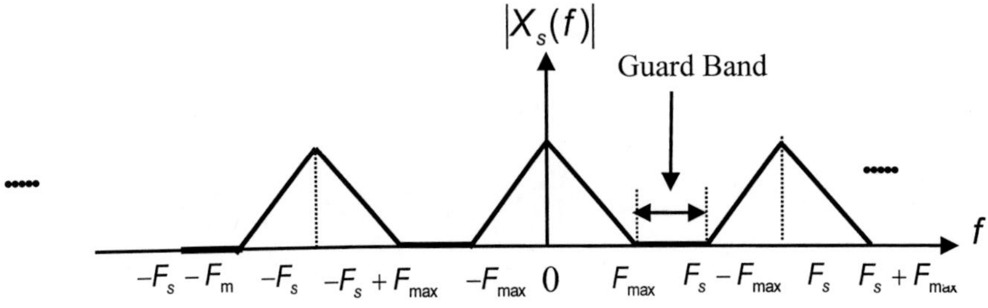

Fig. 1.11. The Fourier transform of the sampled signal with $F_s > 2F_{max}$

where the Guard Band is:

$$G.B = F_{High} - F_{Low}$$
$$= F_s - 2F_{max}$$

In this case, called oversampling, when we sample at a rate which is greater than $2F_{max}$, we say that we are oversampling as shown in Fig.

1.11. All the information contained in the continuous-time signal is found in each of the bands and, in particular, in the band $\left[-F_s, F_s\right]$.

It is, therefore, possible, here, to recover from $X_s(f)$ the continuous signal $X(f)$ isolating in the spectrum of $X_s(f)$ by using a practical Low-Pass Filter (LPF) that removes the sidebands as given in Fig. 1.12.

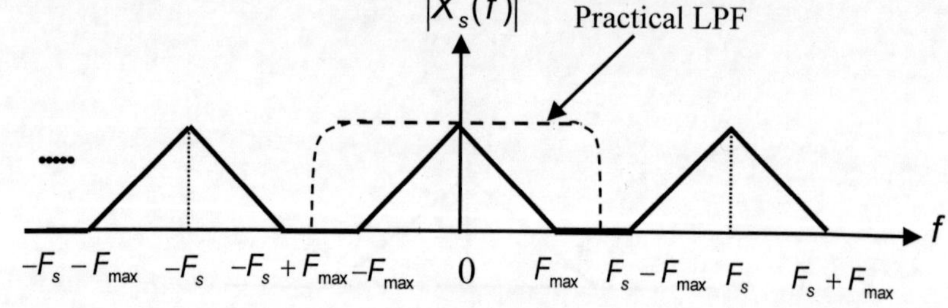

Fig. 1.12. The Fourier transform of the sampled signal with $F_s > 2F_{max}$

Case 2: If $F_s = 2F_{max}$, the Fourier transform of the sampled signal is shown in Fig. 1.13.

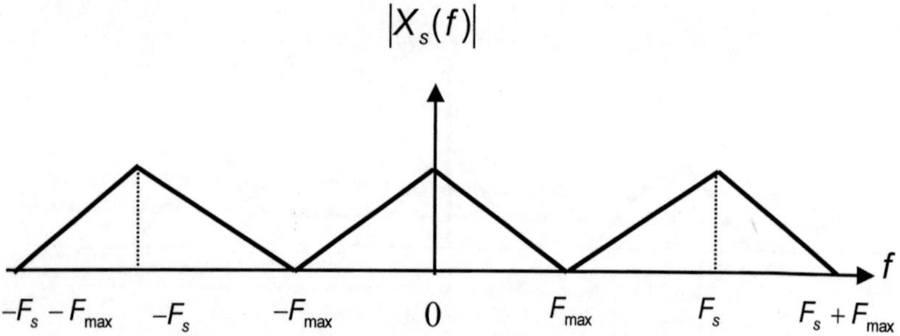

Fig. 1.13. The Fourier transform of the sampled signal with $F_s = 2F_{max}$

In the situation $G.B = 0$, it is possible to reconstruct the base pattern $X(f)$ corresponding to the spectrum of the continuous-time signal $x(t)$ using an ideal low-pass filter with $F_c = F_{max}$, where F_c is the cutoff frequency of the low-pass filter as given in Fig. 1.14.

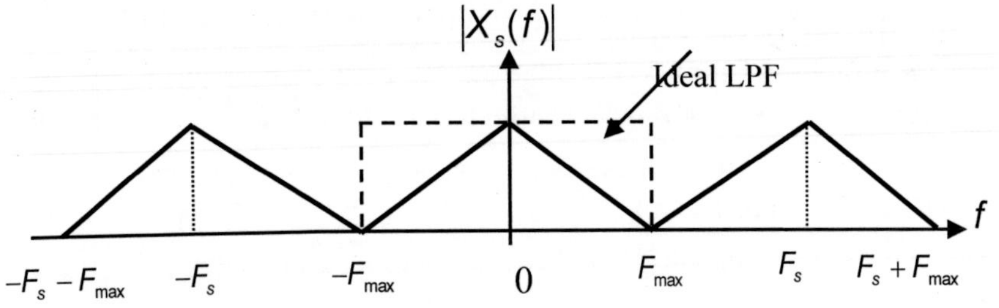

Fig. 1.14. The use of an ideal low-pass filter when $F_s = 2F_{max}$

In this case, it is not recommended to use the practical low-pass filter because the phenomenon of the aliasing effect may be continued as given in the following figure:

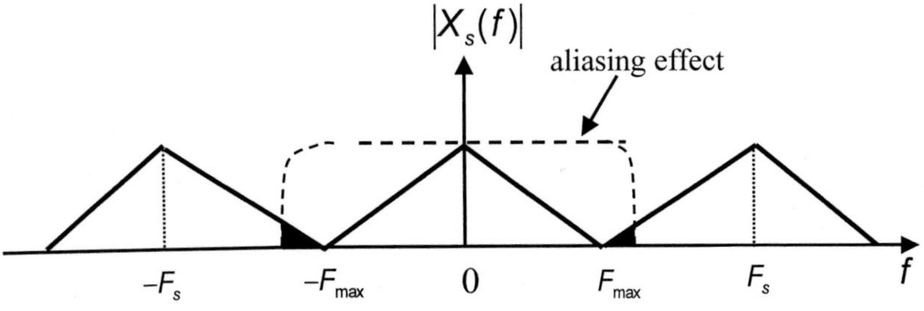

Fig. 1.15. The use of practical low-pass filter when $F_s = 2F_{max}$

Case 3: If $F_s < 2F_{max}$, the Fourier transform of the sampled signal is shown in Fig. 1.16.

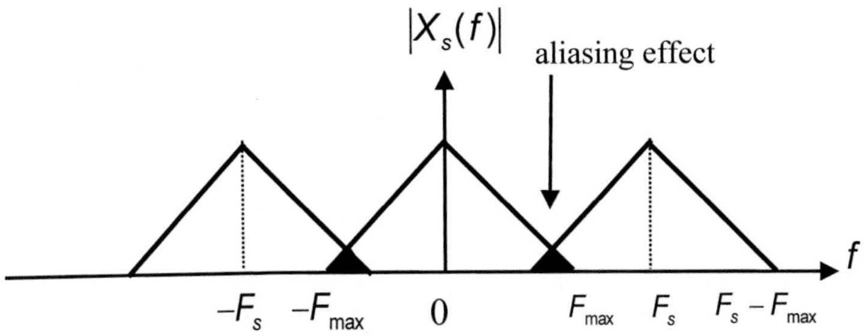

Fig. 1.16. The Fourier transform of the sampled signal with $F_s < 2F_{max}$

In this situation, when we sample at a rate which is less than $2F_{max}$, we say we are undersampling and aliasing will yield misleading results as shown in Fig. 1.16. Distortions occur in the spectrum $\left[-F_s, F_s\right]$, as a result of recombination of its various components. In this case, it is impossible to reconstruct the base pattern $X(f)$ corresponding to the spectrum of the continuous-time signal $x(t)$ even if the practical or ideal low-pass filters are used.

Example:
Let us consider a continuous-time signal $x(t)$ with a maximal frequency $F_{max} = 159.15Hz$, given as:

$$x(t) = 2\sin(1000t) + 3\cos(1000t)$$

The continuous-time signal $x(t)$ is presented by the following figure:

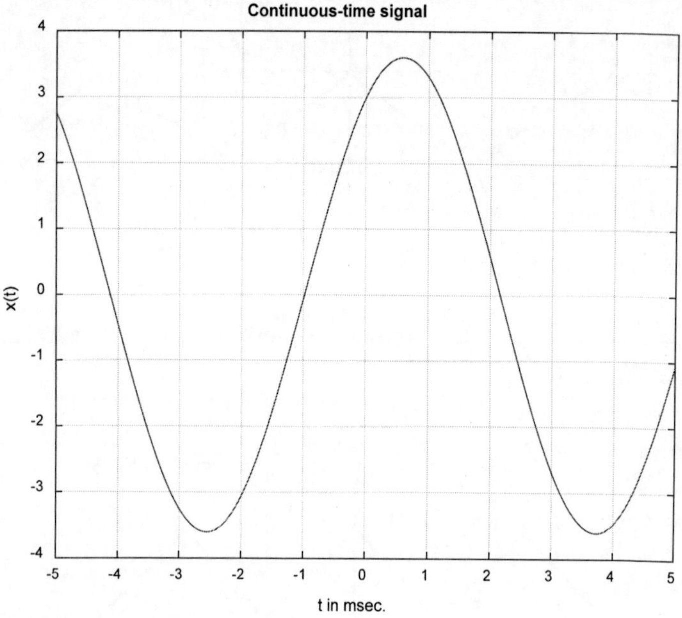

Fig. 1.17. The evolution of $x(t)$

When $F_s > 2F_{max}$, $F_s = 5000Hz$, the discrete-time Fourier Transform $X(f)$ is presented by Fig. 1.18.

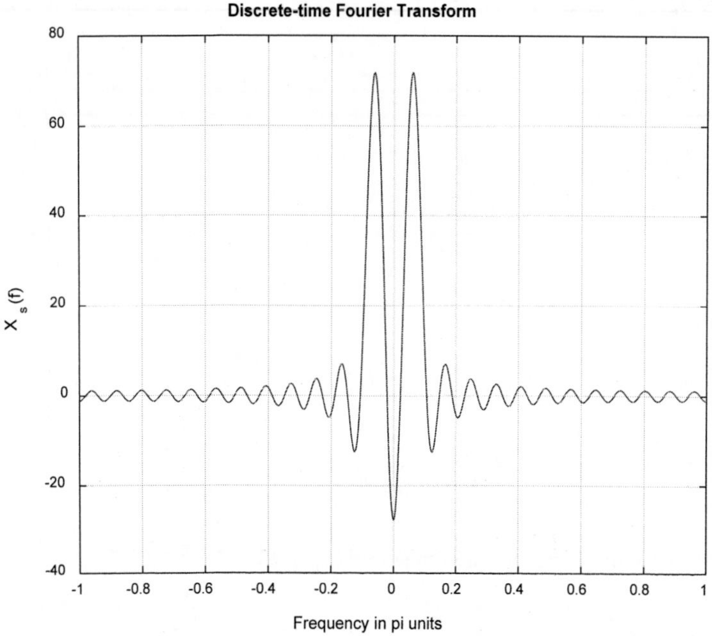

Fig. 1.18. The Fourier transform of the sampled signal with $F_s > 2F_{max}$

If $F_s = 2F_{max} = 318.3098Hz$, the discrete-time Fourier Transform $X_s(f)$ is presented by Fig. 1.19.

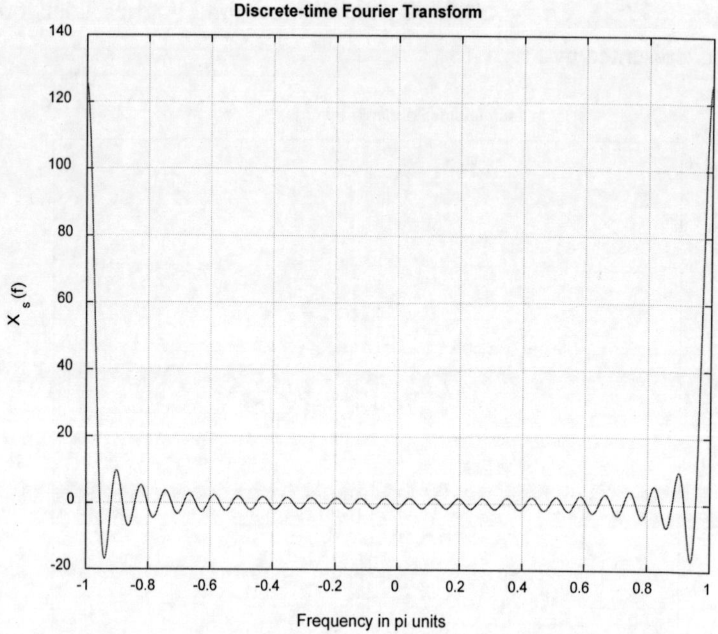

Fig. 1.19. The Fourier transform of the sampled signal with $F_s = 2F_{max}$

When $F_s < 2F_{max}$, $F_s = 1000Hz$, the discrete-time Fourier Transform $X_s(f)$ is presented by the following figure:

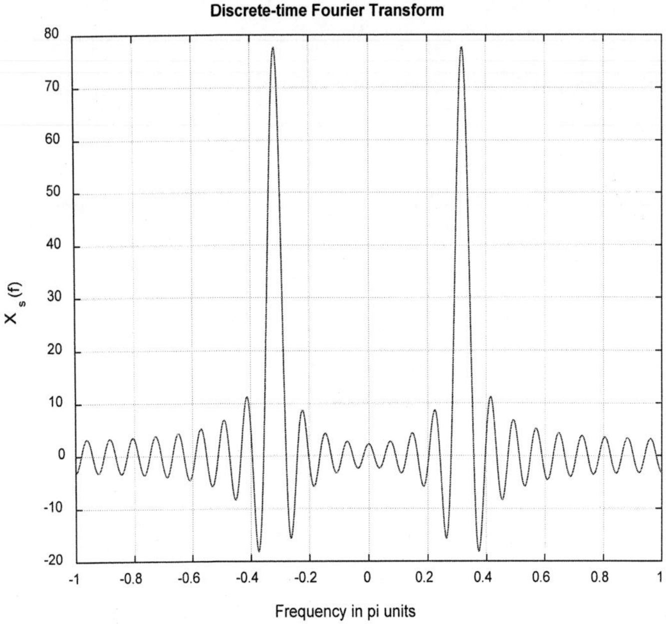

Fig. 1.20. The Fourier transform of the sampled signal with $F_s < 2F_{max}$

b. Anti-aliasing filter

The analog signal, $x(t)$, is sometimes contaminated by high frequency disturbances. Thus, the actual frequency of the signal is greater than the frequency F_{max} that is initially planned without these noises:

$$F_s > 2F_{max}$$

It is compulsory to filter all components of the analog signal whose frequencies are greater than that before the sampling operation F_{max}.

A simple and effective solution is to insert an analog filter called an anti-aliasing filter or a Low-Pass Filter (LPF) before sampling the signal.

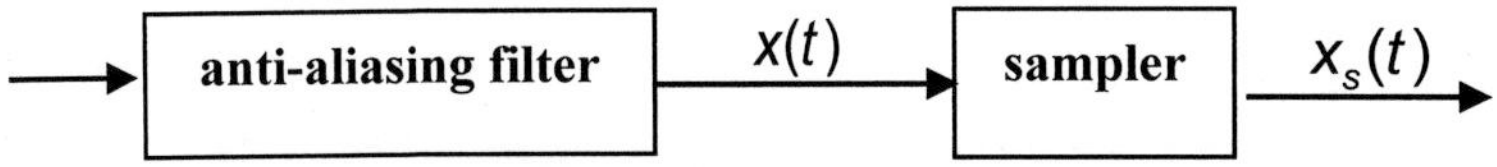

Fig. 1.21. The anti-aliasing filter before the sampler

The used filter must reflect sufficient attenuation at frequencies above $\frac{F_s}{2}$, but, the LPF cutoff frequency must be greater than the closed loop system bandwidth.

The parameters of the low-pass filter are given in the following figure:

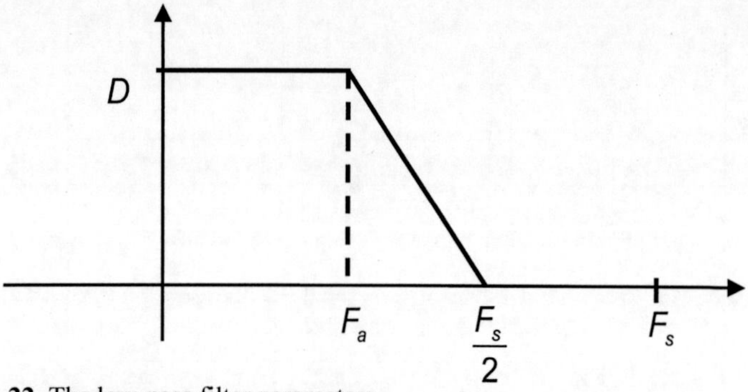

Fig. 1.22. The low-pass filter parameters

These parameters define the transition band characteristics of the filter
F_a : the desired analog input bandwith
F_{pass} : the pass-band of the filter $F_a = F_{pass}$
F_{stop} : the beginning of the filter's stopband $F_{stop} = \frac{F_s}{2}$
D : the desired upper-frequency dynamic range (dB)
M : the approximate order of the filter (the number of poles)

$$M = \frac{D}{6\log_2 \frac{F_s}{2F_a}}$$

c. Nyquist sampling theorem

A signal $x(t)$ which is bandlimited to F_{max} can be completely recovered from its samples if the samples are taken at a rate of

$$F_s \geq 2.F_{max}$$

where T_s is the sampling period, $F_s = \frac{1}{T_s}$ is the sampling rate and $2F_{max}$ is called the minimum sampling rate or the Nyquist rate.

1.2.2. The quantization of a sampled signal

The second aspect of analog-to-digital conversion is quantization. This process converts a discrete in time and continuous in amplitude signal $x_s(t)$ to a signal that is discrete in both time and amplitude $x_q(t)$. The quantization can be either uniform or nonuniform in type. In a uniform quantizer, the representation levels are uniformly spaced; otherwise, the quantizer is nonuniform. In uniform quantization, the quantization regions are chosen to have equal length whereas in nonuniform quantization, regions of variable lengths are used.

1.2.2.1. Uniform quantization

In this section, when the continuous in time signal $x(t)$ is in a finite range $(\min(x), \max(x))$, the length of the quantization region, which is called the quantum, is

$$Q = \frac{\max(x(t)) - \min(x(t))}{L}$$

where $L = 2^N$ is the number of possible states and N is a positive integer or the number of bits which make it possible to express digital signals from 0 to $\left(2^N - 1\right)$ in natural binary code.

Among the methods of uniform quantization, we can find in the literature the rounding method and the truncation method. The rounding method of uniform quantization is given as

$$y = round(z)$$

where the value of the quantization step between levels is

$$z = \frac{x(t) - \min\left(\min(x(t))\right)}{Q}$$

Note that, in comparison with the noise present in the signal, the number of bits chosen should be enough so that the quantization error is small.

Example:
Let us consider the continuous-time signal

$$x(t) = A\sin(wt)$$

with $A = 1$, $w = \frac{2\pi}{T} = 2\pi f$, $T = 1s$ and $f = 1Hz$. The sampling period is $T_s = \frac{1}{F_s} = \frac{1}{3}s$.

The sampled signal of $x(t)$ is given as:

$$x_s(t) = \sum_{k=0}^{+\infty} \sin(2\pi k T_s)\delta(t - kT_s) = \sin(2\pi k T_s) = \sin(\frac{2\pi k}{3})$$

We perform a uniform quantization with $L = 2^2$, $L = 2^3$ and $L = 2^4$ levels to the continuous-time signal $x(t)$ and we plot the output of each one in the same figure with $x(t)$.

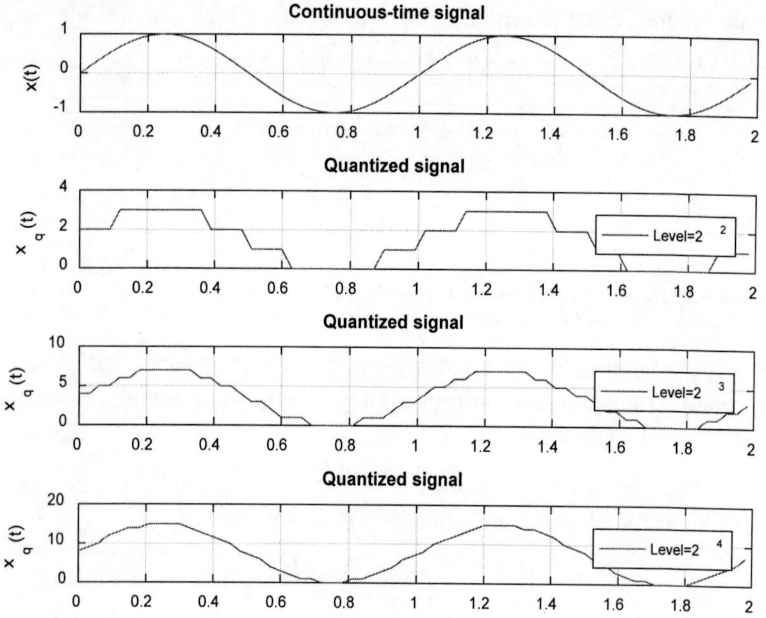

Fig. 1.23. Quantized versions of an analog signal

According to Fig. 1.23, the more quantum levels we use, L=16, the better performance we get. However, among the problems with uniform quantization is that it is only optimal for uniformly a distributed signal, which is why some solutions are proposed that use non-uniform quantization and where the quantization interval is smaller and near zero.

1.2.2.2. Non-uniform quantization

The input of continuous in time signal amplitude distribution is far from being uniformly distributed. Therefore, it makes sense to design a quantizer with less quantization levels at larger amplitudes and more quantization levels at lower amplitudes. The resulting quantizer will be a nonuniform quantizer having variable spacing between the quantization levels.

In nonuniform quantization, the sampled signal is first passed through a nonlinear element that reduces the dynamic range of the signal, then a uniform quantization on the output should be performed.
Two types of compressors are used in nonuniform quantizers; the A-law and the μ-law.

In the A-law quantizer, the input-output relationship is defined by

$$y(x)=\begin{cases}\dfrac{A|x(t)|}{1+\ln(A)}.\text{sgn}(x), & |x(t)|<\dfrac{1}{A}\\[2ex] \dfrac{1+\ln(A|x(t)|)}{1+\ln(A)}.sgn(x), & \dfrac{1}{A}\leq|x(t)|<1\end{cases}$$

In practice, A=87.6.

In the μ-law quantizer, the input-output relation is defined by

$$y(x)=\frac{\log(1+\mu|x(t)|}{\mu}.\text{sgn}(x)$$

where $\mu=255$ and $|x(t)|\leq 1$.

1.2.2.3. Signal-to-quantization-noise-ratio

The signal-to-quantization-noise-ratio (SQNR) in dB is expressed as follows

$$SQNR_{dB} = 10\log\frac{\sigma_x^2}{\sigma_e^2} = 10\log\frac{E(x^2)}{E(x-x_q)^2} = 20\log\frac{|x|}{|x-x_q|}$$

where σ_e^2 is the variance of the quantization noise error and σ_x^2 is the variance of the continuous-time signal $x(t)$.

1.2.3. The coding of quantized signal

After the quantization process comes the third aspect of analog-to-digital conversion which is coding. The coding process converts the quantized level of each sample into bits. The encoding scheme that is usually employed is natural binary coding, meaning that the highest level is mapped into a sequence of all ones and the lowest quantization level is mapped into a sequence of all zeros.

A binary word is written: $b_1\ b_2\ \dots\ b_{N-1}\ b_N$, where b_1 is the most significant bit and b_N is the least significant bit. The corresponding decimal number is:

$$D = b_1 2^{N-1} + b_2 2^{N-2} + \dots + b_{N-1} 2^1 + b_N 2^0$$

Example:
Let us consider the continuous in time signal

$$x(t) = A\sin(wt)$$

with $A=1$, $w=\frac{2\pi}{T}=2\pi f$, $T=1s$, $f=1Hz$ and the sampling period $T_s=\frac{1}{F_s}=\frac{1}{3}s$.

The sampled signal of $x(t)$ is given as:

$$x_s(t) = \sin(\frac{2\pi k}{3})$$

We perform an encoding of this continuous-time signal using differents levels ($L = 2^2$, $L = 2^3$ and $L = 2^4$) as given by the following figure:

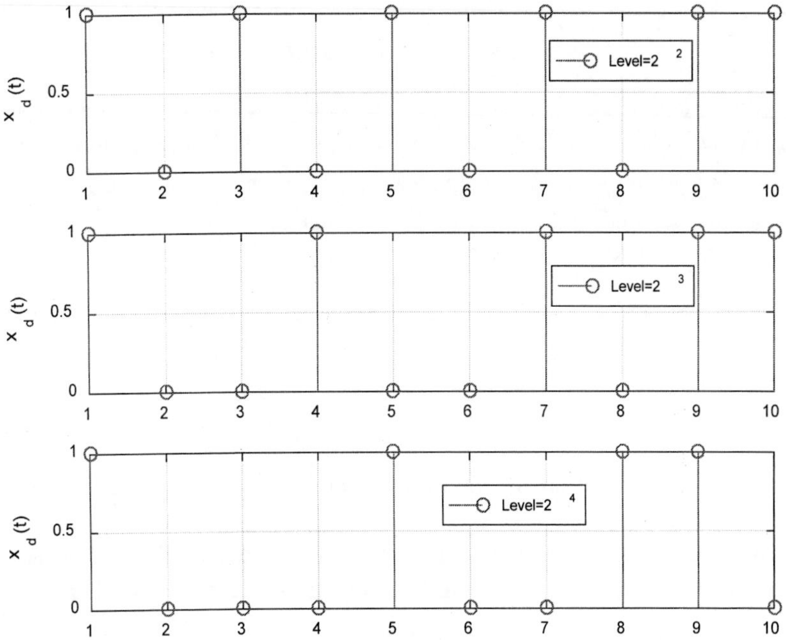

Fig. 1.24. The encoding process of the quantized signal with different levels

Finally, the general process of the signal's digitization is given as:

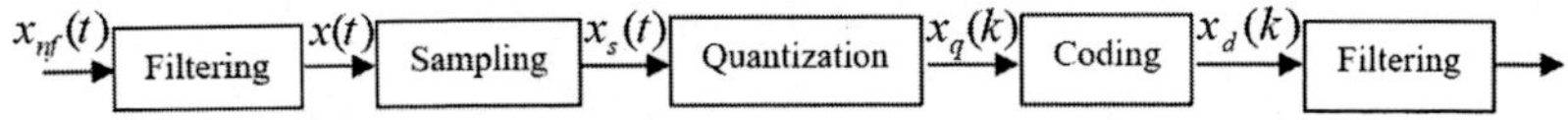

Fig. 1.25. Three processes of digital processing

where $x(t)$ is an analog signal, $x_s(t)$ is a sampled signal, $x_q(k)$ is a quantified signal, $x_d(k)$ is a digital signal and $x_{nf}(t)$ is a non-filtered analog signal.

Filtering is used to comply with the sampling theorem. The filter placed before the sampling is an anti-aliasing filter. It removes frequencies that are higher than half of the sampling rate. The filter placed after the coding is a reconstruction filter. It may include a correction for the method of the hold.

1.3. Reconstruction of analog signal

In the previous section, we discussed the process of sampling, i.e., obtaining a discrete in time signal $x_s(t)$ from a continuous in time signal $x(t)$. But we now consider the reverse problem, namely how to construct a continuous-time signal given samples that are discrete in time. This operation is called reconstruction:

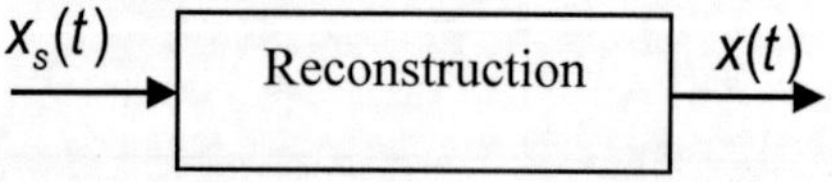

Fig. 1.26. The process of reconstruction signal

where $x(t)$ is a continuous in time signal and $x_s(t)$ is a sampled signal.

In literature, the reconstruction process is found by the data hold. This latter is a process of generating a continuous-time signal $h(t)$ from a sampled signal $x_s(t)$. During the time interval $kT_s \le t \le (k+1)T_s$, the signal $h(t)$ may be approximated by a polynomial that depends on α or in terms of:

$$h(kT_s + \alpha) = a_0 + a_1\alpha + ... + a_{n-1}\alpha^{n-1} + a_n\alpha^n$$

with $0 \le \alpha \le T_s$ and we note $h(kT_s) = x(kT_s)$, so

$$h(kT_s + \alpha) = x(kT_s) + a_1\alpha + ... + a_{n-1}\alpha^{n-1} + a_n\alpha^n$$

To generate $h(kT_s + \alpha)$, the nth-order hold uses the past $(n+1)$ discrete data $x((k-n)T_s)$, $x((k-n+1)T_s)$, ..., $x(kT_s)$.

1.3.1. Zero-Order-Hold

The Zero-Order-Hold (ZOH) is found when $n = 0$ in the above equation, so that:

$$h(kT_s + \alpha) = x(kT_s) \quad 0 \le \alpha \le T_s \text{ and } k = 0, 1, 2, ...$$

In order to find the transfer function of the ZOH, let us consider $u(t)$ as a unit step function input, as given in this case:

$$h(t) = x(0)[u(t) - u(t - T_s)] + x(T_s)[u(t - T_s) - u(t - 2T_s)] + x(2T_s)[u(t - 2T_s) - u(t - 3T_s)] + \dots$$

$$= \sum_{k=0}^{+\infty} x(kT_s)[u(t - kT_s) - u(t - (k+1)T_s)]$$

and suppose that

$$L[u(t - kT_s)] = \frac{e^{-kT_s}}{s}$$

Thus

$$L[h(t)] = H(s)$$

$$= \sum_{k=0}^{+\infty} x(kT_s) \frac{e^{-kT_s} - e^{-(k+1)T_s}}{s}$$

$$= \frac{1 - e^{-T_s}}{s} \sum_{k=0}^{+\infty} x(kT_s) e^{-kT_s}$$

$$= H_0(s) \sum_{k=0}^{+\infty} x(kT_s) e^{-kT_s}$$

Then, the transfer function of ZOH is

$$H_0(s) = \frac{H(s)}{\sum_{k=0}^{+\infty} x(kT_s) e^{-kT_s}}$$

$$= \frac{1 - e^{-sT_s}}{s}$$

The Input-Output of the ZOH process is given by Fig. 1.27 as:

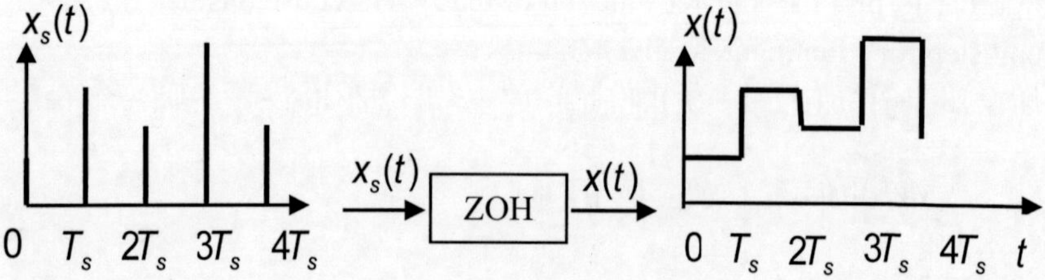

Fig. 1.27. The process of reconstruction signal by ZOH

Example:
Using Simulink/MatLab, in the following example, a unit step signal is used. Two sampling periods are taken, ($T_s = 0.1s$ and $T_s = 1s$), by applying the zero-order-hold.

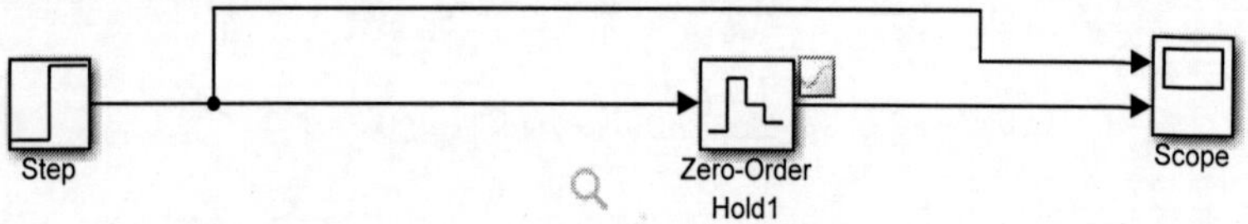

Fig. 1.28. The Zero-Order-Hold by Simulink/MatLab

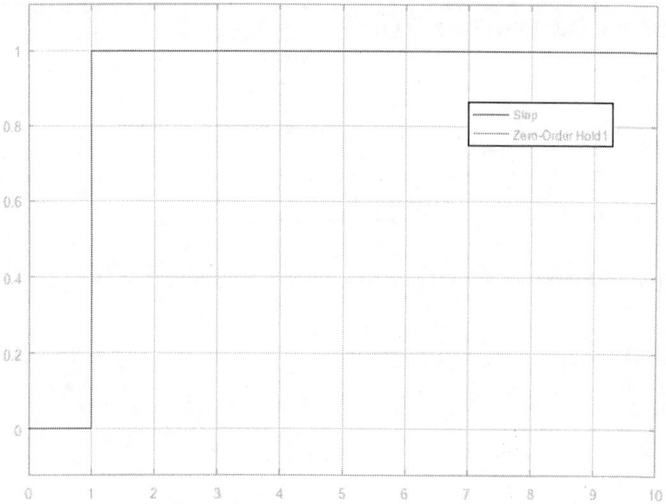

Fig. 1.29. The input and output from ZOH with $T_s = 0.1s$

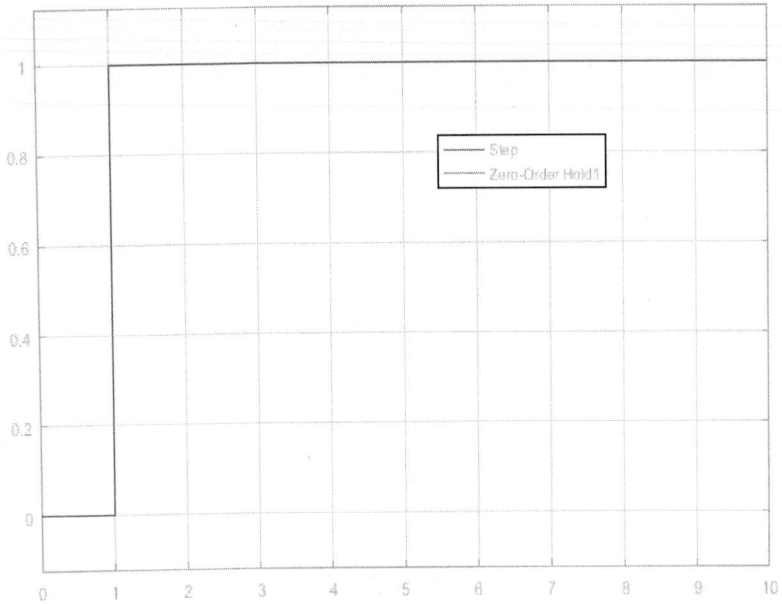

Fig. 1.30. The input and output from ZOH with $T_s = 1s$

Example:
To show more of the influence of the ZOH, in the following example, a sinusoidal signal is treated using two sampling periods, $T_s = 0.1s$ and $T_s = 1s$, by applying the zero-order-hold.

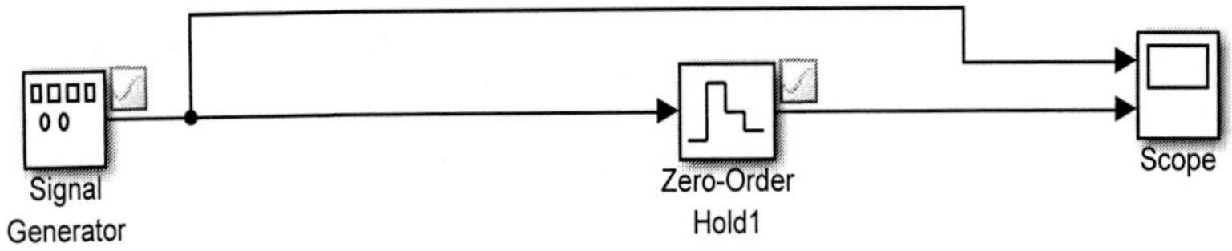

Fig. 1.31. The Zero-Order-Hold by Simulink/MatLab

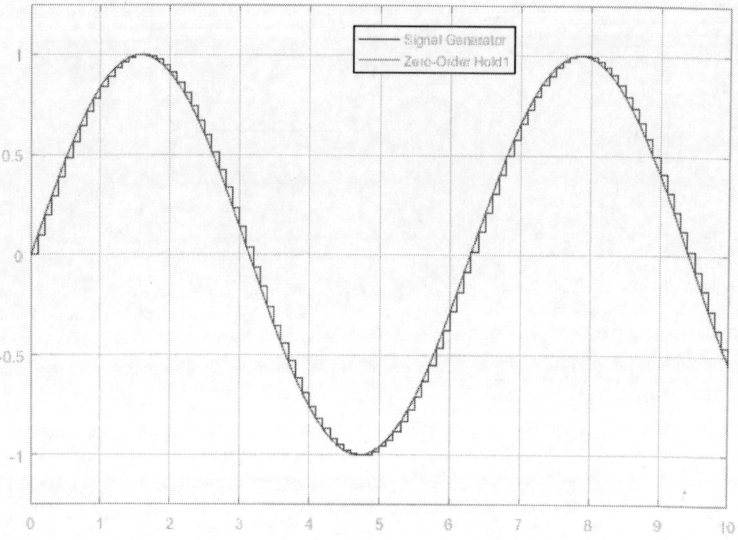

Fig. 1.32. The input and output from ZOH with $T_s = 0.1s$

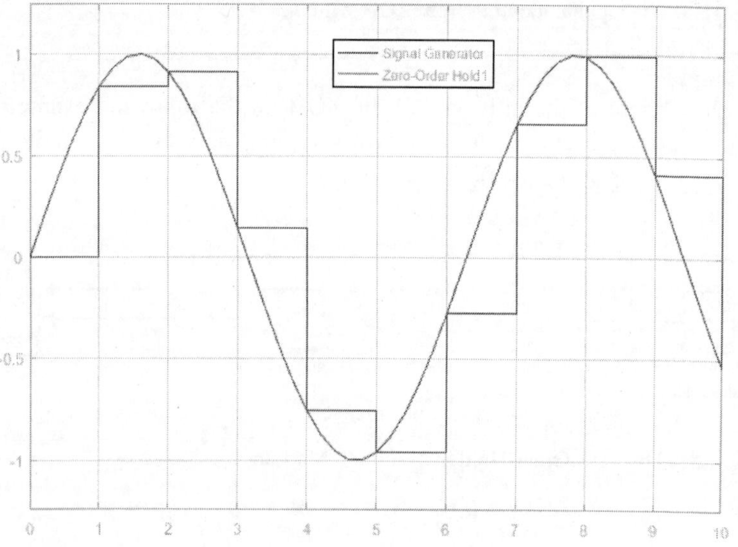

Fig. 1.33. The input and output from ZOH with $T_s = 1s$

Using the zero-order-hold, in the case where $T_s = 0.1s$, the reconstruction is better than in the case where the sampling period $T_s = 1s$.

1.3.2. First-Order Hold

The First-Order-Hold (FOH) is found when $n = 1$ in the above equation, so that:

$$h(kT_s + \alpha) = a_1\alpha + x(kT_s) \quad 0 \le \alpha \le T_s \text{ and } k = 0, 1, 2 \dots$$

Now

$$h((k-1)T_s) = x((k-1)T_s)$$

so that

$$h((k-1)T_s) = -a_1T_s + x(kT_s)$$
$$= x((k-1)T_s)$$

or

$$a_1 = \frac{x(kT_s) - x((k-1)T_s)}{T_s}$$

then

$$h(kT_s + \alpha) = \frac{x(kT_s) - x((k-1)T_s)}{T_s}\alpha + x(kT_s) \text{ , } 0 \le \alpha \le T_s \text{ and }$$
$$k = 0, 1, 2 \dots$$

To find the transfer function of the FOH, let us consider $u_s(t)$ as a unit step function input as given in the following equation:

$$u_s(t) = \sum_{k=0}^{+\infty} u(kT_s)\delta(t - kT_s) = \sum_{k=0}^{+\infty} \delta(t - kT_s)$$

In this case:

$$h(t) = (1 + \frac{t}{T_s})u(t) - \frac{t - T_s}{T_s}u(t - T_s) - u(t - T_s)$$

Thus, the transfer function is:

$$H(s) = \left(\frac{1}{s} + \frac{1}{s^2 T_s}\right) - \frac{1}{s^2 T_s} e^{-sT_s} - \frac{1}{sT_s} e^{-sT_s}$$

$$= \frac{1 - e^{-sT_s}}{s} + \frac{1 - e^{-sT_s}}{s^2 T_s}$$

$$= \left(1 - e^{-sT_s}\right) \frac{1 + sT_s}{s^2 T_s}$$

The Laplace transform of the unit step is:

$$X_s(s) = L\left[u_s(t)\right] = \frac{1}{1 - e^{-sT_s}}$$

so the transfer function becomes:

$$H(s) = H_1(s) \frac{1}{1 - e^{-sT_s}}$$

$$= \frac{1 + sT_s}{T_s} \left(\frac{1 - e^{-sT_s}}{s}\right)^2$$

Thus, the transfer function of FOH is:

$$H_1(s) = \frac{H(s)}{\left(\frac{1}{1 - e^{-sT_s}}\right)} = \frac{1 + sT_s}{T_s} \left(\frac{1 - e^{-sT_s}}{s}\right)^2$$

The Input-Output of the FOH process is presented as given:

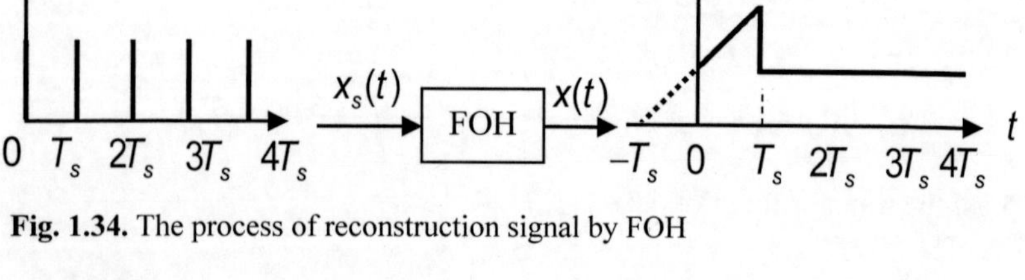

Fig. 1.34. The process of reconstruction signal by FOH

Example:

Applying Simulink/MatLab, the response of the FOH when the input is a unit step is given by the following figure:

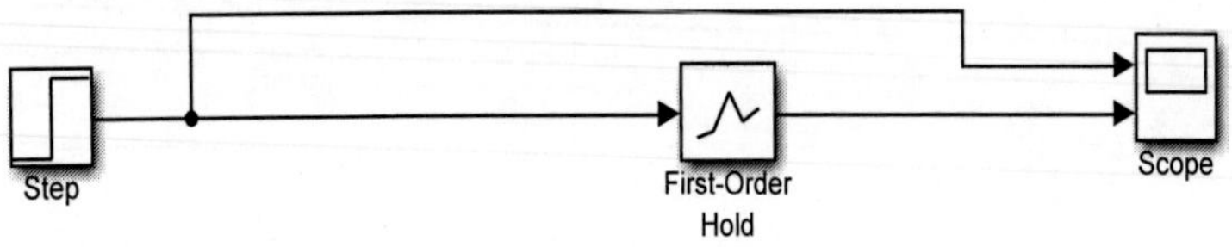

Fig. 1.35. The First-Order-Hold by Simulink/MatLab

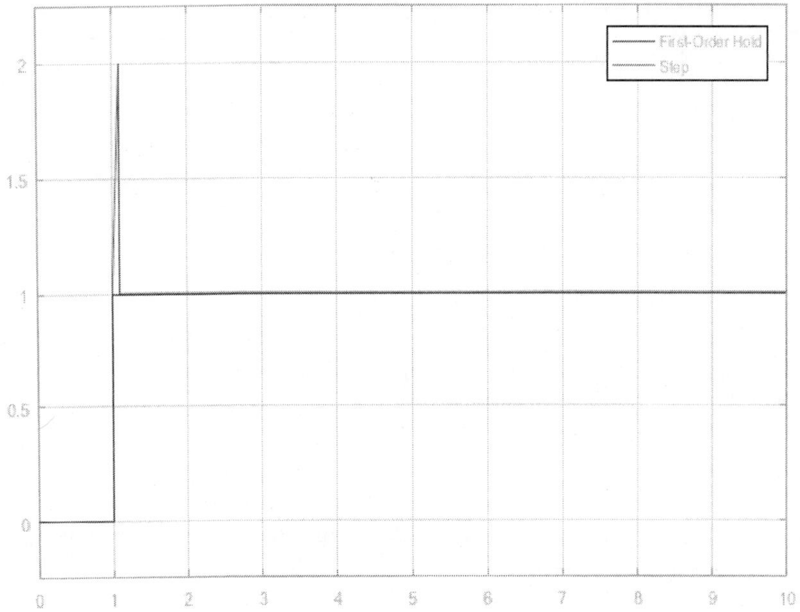

Fig. 1.36. The input and output from FOH with $T_s = 0.1s$

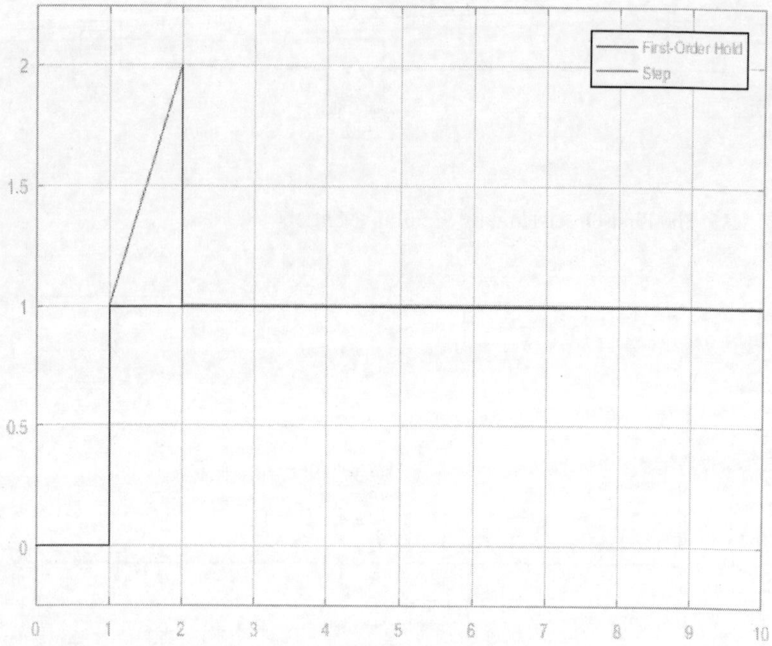

Fig. 1.37. The input and output from FOH with $T_s = 1s$

Example:
In the following example, we take a sinusoidal signal and we use two sampling periods; the first $T_s = 0.1s$ and the second $T_s = 1s$ by applying the first-order-hold.

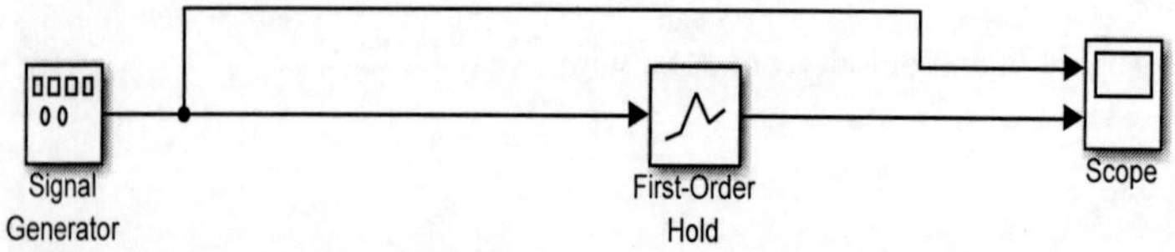

Fig. 1.38. The First-Order-Hold by Simulink/MatLab

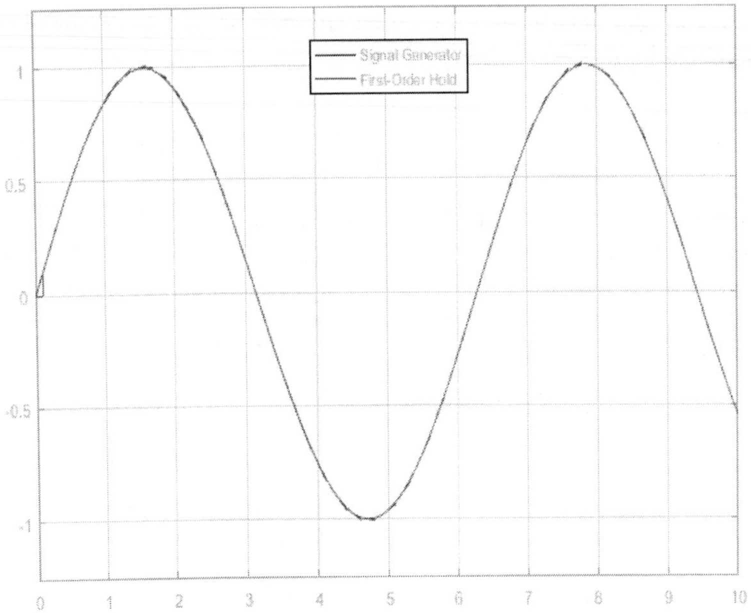

Fig. 1.39. The input and output from FOH with $T_s = 0.1s$

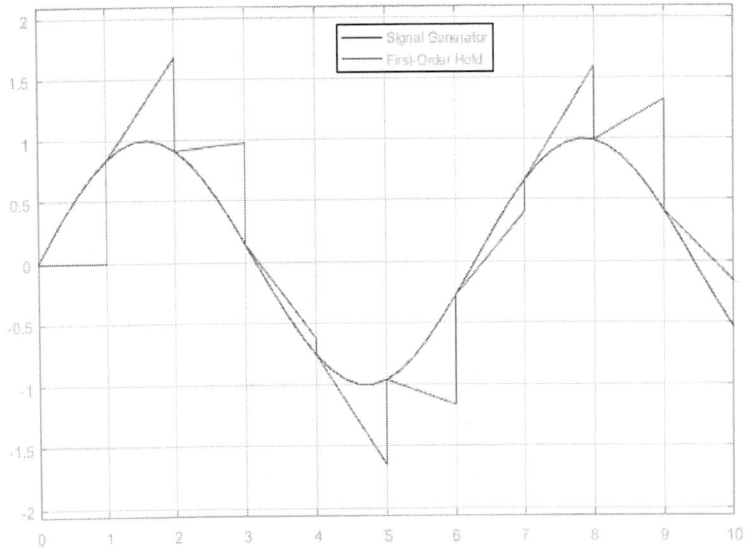

Fig. 1.40. The input and output from FOH with $T_s = 1s$

Using the first-order-hold, in the case $T_s = 0.1s$, the reconstruction is better than it was in the case where the sampling period $T_s = 1s$.

We conclude that, the first-order-hold gives a better result than that the one obtained with the zero-order-hold.

The zero-order-hold interpolation is one of the most widely used methods and it is easily implemented. Another conceptually simple method is the linear interpolation, which is also called the first-order-hold interpolation. With this method, the reconstruction is a continuous function that just connects the sample values with straight lines. Higher order interpolation schemes also exist that pass smoother functions through the samples. The higher the order is, the more samples are needed to be used to reconstruct a value at each time.

1.3.3. Cubic-order-hold interpolation

This approach, for a smoother reconstruction, uses spline interpolation, but it is not necessarily a more accurate estimate of the analog signal between samples. Hence this interpolation does not require an analog post-filter. By using a set of piecewise continuous third-order polynomials called cubic splines, the smoother reconstruction is obtained.

Example: by applying the cubic-order-hold interpolation, the reconstruction of the analog signal from the discrete-time signal $x(k) = \sin(2\pi kf)$ is given by the following figure using the spline MatLab function.

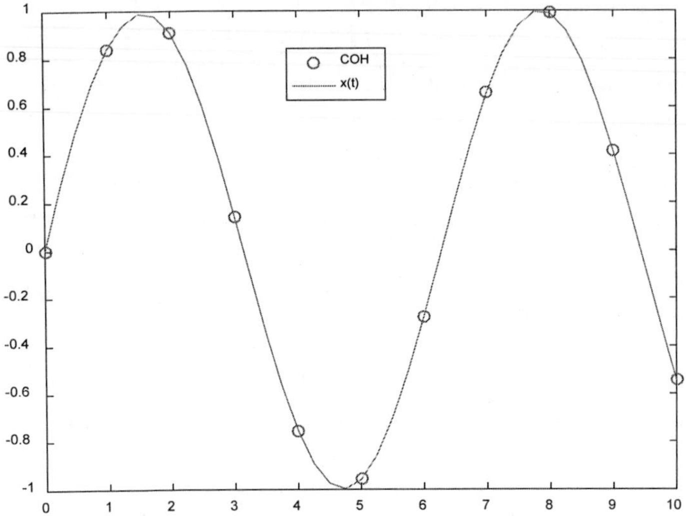

Fig. 1.41. The discrete-time signal and the analog by COH method

1.4. Conclusions

In this chapter, emphasis has been placed on digitizing an analog signal. Indeed, sampling, quantization and coding are well detailed. In the second part of this chapter the reconstruction of a signal is proved. Some application exercises are corrected to ensure the basic notions already defined.

1.5. Applications

Application 1

1 - The signal $x_1(t)=\sin(2\pi t)$, $t\geq 0$, is sampled, where the sampling rate $F_s=3Hz$. Find the sampled signal $x_{1s}(t)$ and draw it.

2 - Perform the same operation with the signal $x_2(t)=\sin(2\pi t)+\sin(6\pi t)$, $t\geq 0$. Find the sampled signal $x_{2s}(t)$ and draw it.

3 - Compare these sampled signals, $x_{1s}(t)$ and $x_{2s}(t)$.

Answer

1- The signal $x_1(t)=\sin(2\pi t)=\sin(w_1 t)$ with $w_1=\frac{2\pi}{T_1}=2\pi f_1$, $T_1=1s$ and $f_1=1Hz$. The sampling period is $T_s=\frac{1}{F_s}=\frac{1}{3}s$. The sampled signal of $x_1(t)$ is $x_{1s}(t)$:

$$x_{1s}(t)=\sum_{k=0}^{+\infty}\sin(2\pi kT_s)\delta(t-kT_s)=\sin(2\pi kT_s)=\sin(\frac{2\pi k}{3})$$

$$\forall k\geq 0$$

The analog signal $x_1(t)$ and the sampled signal $x_{1s}(t)$ are presented in the following figure:

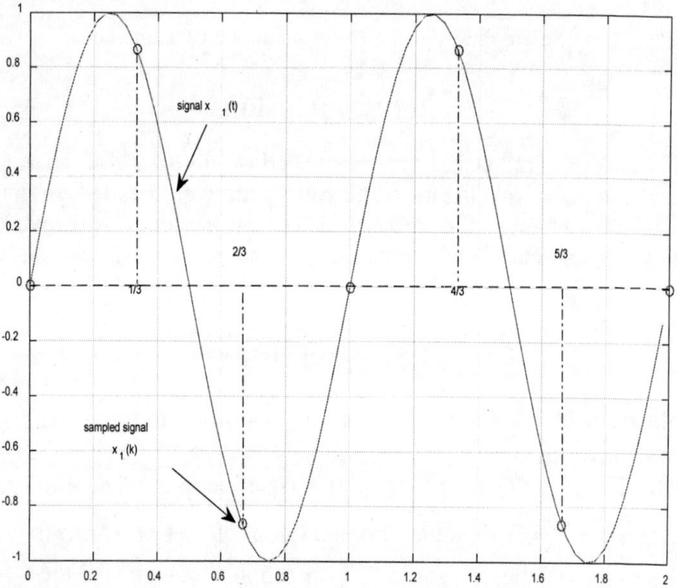

Fig. 1.42. The analog signal $x_1(t)$ and the sampled signal $x_{1s}(t)$

2- The sampled signal of $x_2(t)$ is $x_{2s}(t)$:

$$x_{2s}(t) = \sum_{k=0}^{+\infty} (\sin(2\pi k T_s) + \sin(6\pi k T_s))\delta(t - kT_s)$$

$$= \sum_{k=0}^{+\infty} \sin(2\pi k T_s)\delta(t - kT_s) + \sum_{k=0}^{+\infty} \sin(6\pi k T_s)\delta(t - kT_s)$$

$$= \sin(2\pi k T_s) + \sin(6\pi k T_s) = \sin(\frac{2\pi k}{3}) + \sin(\frac{6\pi k}{3}) = \sin(\frac{2\pi k}{3})$$

The analog signal $x_2(t)$ and the sampled signal $x_{2s}(t)$ are presented in the following figure:

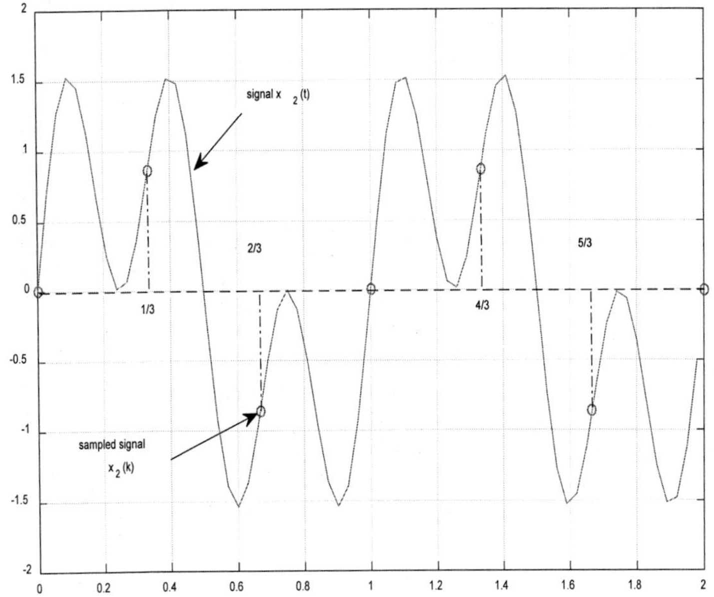

Fig. 1.43. The analog signal $x_2(t)$ and the sampled signal $x_{2s}(t)$

3 - From questions 1 and 2, we find:

$$x_{2s}(t) = x_{1s}(t)$$

This example illustrates that two sampled sinusoids can produce the same discrete in time signal. When this occurs, we say that these signals are aliases of each other.

The signal $x_2(t) = \sin(2\pi t) + \sin(6\pi t) = \sin(w_1 t) + \sin(w_2 t)$, with

$$w_1 = \frac{2\pi}{T_1} = 2\pi f_1 = 2\pi\ ,\ T_1 = 1s\ \ f_1 = 1Hz\,;$$

$$w_2 = \frac{2\pi}{T_2} = 2\pi f_2 = 6\pi\ ,\ T_2 = \frac{1}{3}s \text{ and } f_2 = 3Hz\,.$$

We did not respect the conditions stated in the Nyquist theorem that are related to $x_{2s}(t)$. Indeed, $x_2(t)$ has two frequencies, $f_1 = 1Hz$ and $f_2 = 3Hz$. The maximum frequency is clear $F_{\max} = f_2 = 3Hz$ which is not strictly inferior to $\frac{F_s}{2}$. The sampling of $\sin(6\pi t)$ gives a null signal; in this case, sampling at frequency $F_s = 3Hz$ masks this oscillation. A continuous in time signal $x(t)$ with frequencies no higher than $F_{\max}$ can be reconstructed exactly from its samples.

$x_2(t)$ illustrates that two sampled signals can produce the same discrete in time signal so these signals are aliases of each other.

Application 2

We consider the analog signal $x(t) = \cos(100\pi t) - 2\sin(400\pi t)$ $\forall t \geq 0$.

1- From what frequency can we sample this signal if we want to respect the sampling theorem?

2- We sample $x(t)$ at the sampling rate $F_s = 300Hz$.

  2-a- What should we observe?

  2-b- Determine the sampled signal of $x(t)$.

  2-c- Check the result of 2-a.

Answer

We consider the analog signal $x(t) = \cos(100\pi t) - 2\sin(400\pi t)$ $\forall t \geq 0$.

1- The analog signal is

$$x(t) = \cos(100\pi t) - 2\sin(400\pi t)$$
$$= \cos(w_1 t) - 2\sin(w_2 t)$$

with $w_1 = \frac{2\pi}{T_1} = 2\pi f_1 = 100\pi$ so $T_1 = \frac{1}{50}s$ and $f_1 = 50Hz$;

$w_2 = \frac{2\pi}{T_2} = 2\pi f_2 = 400\pi$ so $T_2 = \frac{1}{200}s$ and $f_2 = 200Hz$.

The maximum frequency is $F_{\max} = f_2 = 200Hz$.

To respect the conditions stated in the sampling theorem, it is necessary that $F_s > 2F_{\max} = 2f_2 = 400Hz$.

2- If we sample $x(t)$ with the given sampling rate $F_s = 300Hz$, this means $T_s = \frac{1}{F_s} = \frac{1}{300}s$.

2-a- We have $F_s < 400Hz$ so the sampling theorem conditions are not respected, and we will observe a phenomenon of aliasing which generates losses of information.

2-b- The sampled signal of $x(t)$ is $x_s(t)$:

$$x_s(t) = \sum_{k=0}^{+\infty}(\cos(100\pi kT_s) - 2s\text{in}(400\pi kT_s))\delta(t - kT_s)$$

$$= \sum_{k=0}^{+\infty}\cos(100\pi kT_s)\delta(t - kT_s) - \sum_{k=0}^{+\infty}2\sin(400\pi kT_s)\delta(t - kT_s)$$

$$= co\text{s}(100\pi kT_s) - 2\sin(400\pi kT_s) = \cos(\frac{\pi k}{3}) - 2\sin(\frac{4\pi k}{3})$$

$$= \cos(\frac{\pi k}{3}) + 2\sin(\frac{2\pi k}{3})$$

2-c- We have

$$\frac{2\pi k}{3}=\frac{2\pi k\times 100}{300}=\frac{200\pi k}{F_s}=200\pi kT_s=200\pi t=2\times F_{\max}$$ so the result of 2-a holds true.

Application 3

We sample the output of a first order linear system $y(t)=\alpha(1-e^{-\frac{t}{\tau}})$, $t\geq 0$, with a sampling period T_s. α and τ are respectively the static gain and the constant of time.

1- Determine the time t_1 from which $y(t)$ is less than 10% of the steady state value.

2- Find the sampled signal $y_s(t)$. Compare the steady state values of $y(t)$ and $y_s(t)$.

3- For the different values of $T_s=\frac{\tau}{2}$, τ and 2τ:

a) Calculate the first 10 values of $y_s(t)$ according to α.

b) Deduce the rank n from which all samples become less than 10% of the steady state value. Specify the corresponding time and compare to t_1.

4- Deduce the maximum value of the ratio $\frac{T_s}{\tau}$ which makes it possible to sample the analog signal $y(t)$ without losing too much information on the transient regime.

Answer

1 - The value of the steady state of $y(t)$ is

$$y(+\infty)=\lim_{t\to+\infty}y(t) \quad \Rightarrow \quad y(\infty)=\lim_{t\to\infty}\alpha(1-e^{-\frac{t}{\tau}})=\alpha$$

The time from which $y(t)$ is less than 10% of the steady state is:

$$y(t_1)=\lim_{t\to t_1}\alpha(1-e^{-\frac{t}{\tau}})=\alpha(1-e^{-\frac{t_1}{\tau}})=0.9y(\infty)\Rightarrow t_1=2.3\tau\ .$$

2- The sampled signal is

$$y_s(t)=\sum_{k=0}^{+\infty}\alpha(1-e^{-\frac{kT_s}{\tau}})\delta(t-kT_s)=\alpha(1-e^{-\frac{kT_s}{\tau}})$$

The value of the steady state of $y_s(t)$ is

$$\lim_{k\to\infty}y_s(t)=\lim_{k\to\infty}\alpha(1-e^{-\frac{kT_s}{\tau}})=\alpha \text{ so } y(\infty)=y_s(\infty)=\alpha\ .$$

3- For different values of $T_s=\frac{\tau}{2}$, τ and 2τ we have

a- The first 10 values of $y_s(t)$ according to α are summarized in the following table:

Table 1.1. The first 10 values of $y_s(t)$ according to α

	y(0)	y(T$_s$)	y(2T$_s$)	y(3T$_s$)	y(4T$_s$)	y(5T$_s$)	y(6T$_s$)	y(7T$_s$)	y(8T$_s$)	y(9T$_s$)
$T_s=\frac{\tau}{2}$	0	0.39 α	0.63 α	0.78 α	0.86 α	0.92 α	0.95 α	0.97 α	0.99 α	0.99 α
$T_s=\tau$	0	0.63 α	0.86 α	0.95 α	0.98 α	0.99 α	0.998 α	0.99 α	α	α
$T_s=2\tau$	0	0.86 α	0.98 α	0.99 α	α	α	α	α	α	α

b- The samples which are less than 10% of the steady state are obtained if $y(nT_s)\geq y(+\infty)$

- If $T_s=\frac{\tau}{2}$ then the rank n is equal to 5 because $y(5T_s)=0.92\alpha\geq 0.9\alpha$, so the corresponding time is $5T_s=5\frac{\tau}{2}=2.5\tau\geq t_1$

- If $T_s = \tau$ then the rank n is equal to 3 because $y(3T_s) = 0.95\alpha \geq 0.9\alpha$, so the corresponding time is $3T_s = 3\tau \geq t_1$
- If $T_s = 2\tau$ the rank n is equal to 2 because $y(2T_s) = 0.98\alpha \geq 0.9\alpha$, so the corresponding time is $2T_s = 4T \geq t_1$ $2T_s = 4\tau \geq t_1$

4- From $n = 2$, we have $T_s = 2\tau$ which makes it possible to sample the analog signal $y(t)$ every two time constants. We are losing a lot of information about the transitional regime, i.e., $\frac{T_s}{\tau} = 2$.

From $n = 3$, we have $T_s = \tau$ which makes it possible to sample the analog signal $y(t)$ for each T_s. We have more information on the transitional regime, that is $\frac{T_s}{\tau} = 1$.

The maximum value of the ratio $\frac{T_s}{\tau}$ is 1 while for a first order system we have $\frac{\tau}{4} < T_s < \tau$.

Application 4

Use the ZOH method, the FOH method and the COH method, and find the original signal:

$$x(t) = \sin(t).$$

using the sampling rate $F_s = 450Hz$.

Answer

In order to recover the original signal, we used the stairs function for the ZOH method, plot function for the FOH method and spline function for the COH method as given in the following figure.

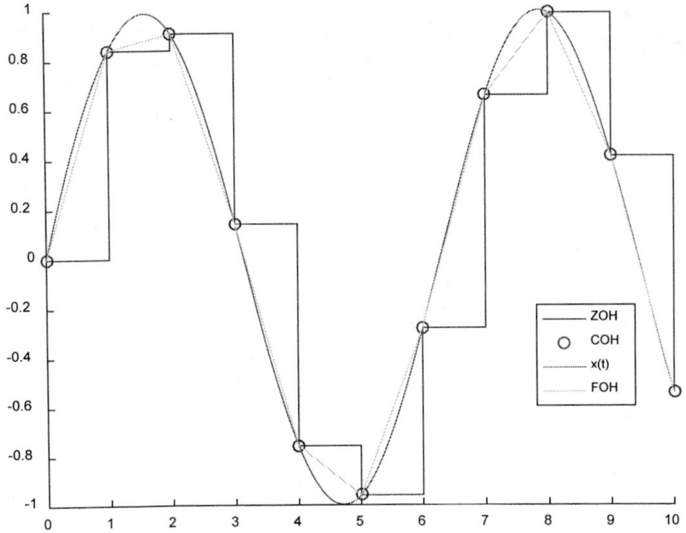

Fig. 1.44. The original signal, the ZOH method, the FOH method and the COH method

1.6 References

Bennett, W.R. 1948. Spectra of quantized signals. Bell System Technical Journal. N. 27: 446-472.

Benenvuto, N., et al., 1986. The 32-kbps ADPCM coding standard, AT & T Technical Journal, Vol. 65, N.5: 12-22.

Zhengtao, D. 2013. Nonlinear and adaptive control systems? The Institution of Engineering and Technology.

Caves, J.T., et al. Sampled analog filtering using switched capacitors as resistor equivalents. IEEE J. Solid State Circuits, Vol. SC-12: 592-599.

Dan, S. 1986. Analog to digital conversion handbook, Third Edition, Prentice-Hall.

Decina, M. and Modena, G. 1988. CCITT standards on digital speech processing, IEEE Journal on Selected Areas of Communication, Vol. 6: 227-234.

Frederic, J.H. 1978. On the use of windows for harmonic analysis with the discrete Fourier transform. IEEE Journal Proceedings, Vol. 66, N. 1: 51-83.

Gray, G.A. and Zeoli, G.W. 1971. Quantization and saturation noise due to analog digital conversion. IEEE Transactions on Aerospace and Electronic Systems: 222-223.

Jayant, N.S and Noll, P. 1984. Digital coding of waveforms: principles and applications to speech and video, Prentice Hall.

Doernberg, J., Lee, H.-S, and Hodges, D.A., 1984. Full speed testing of A/D converters. IEEE Journal of Solid-State Circuits, Vol. SC-19, N. 6: 820-827.

Lawrence, R. and Bernard, G. 1975. Theory and application of digital signal processing. Prentice-Hall.

Leon, W. Couch, II. 2001. Digital and analog communication systems, Pearson Education Asia.

Derek, A. 2015. Control engineering problems with solutions: bookboon.com ISBN 978-87-403-0374-2.

Simon, H. 2001. Communication systems, (4th ed.) John Wiley.

Oppenheim, A.V and Schafer, R.W. 1989. Discrete-time signal processing. Englewood Cliffs, NJ: Prentice Hall.

Oppenheim, A.V, and Willsky, A.S., Syed, H.N. 1997. Signals and systems. Prentice Hall, Upper Saddle River, NJ.

Oppenheim, A.V., Schafer, R.W., and Buck, J.R. 2000. Discrete-time signal processing, (2nd ed.), Pearson Education.

Ziemer, R.E, Tranter, W.H. and Fannin, D.R. 1993. Signals and systems: continuous and discrete. New York, NY: Macmillan.

Mulleti, S. and Seelamantula, C.S. *2016.* Sampling and reconstruction of time-limited signals using sum-of-sincs kernel, *2016 Twenty Second National Conference on Communication (NCC),* Guwahati, India, 2016, pp. 1-6, doi: 10.1109/NCC.2016.7561141.

Vetterli, M., Marziliano, P. and Blu, T. 2002. Sampling signals with finite rate of innovation, *IEEE Trans. Signal Process.,* Vol. 50, no. 6, pp. 1417-1428.

Tur, R., Eldar, Y.C. and Friedman, Z. 2011. Innovation rate sampling of pulse streams with application to ultrasound imaging, *IEEE Trans. Signal Process.,* Vol. 59, no. 4: 1827-1842.

Mulleti, S., Nagesh, S., Langoju, R., Patil, A and Seelamantula, C.S. 2014. Ultrasound image reconstruction using the finite-rate-of-innovation principle, *Proc. IEEE Int. Conf. Image Process. (ICIP):* 1728-1732.

Arild, W. 2013. Discrete dynamical systems with an introduction to discrete optimization problems: bookboon.com ISBN 978-87-403-0327-8.

Castorena, J. and Creusere, C. D. 2015. Sampling of time-resolved full-waveform LIDAR signals at sub-Nyquist rates, *IEEE Trans. Geoscience and Remote Sensing,* Vol. 53, no. 7: 3791-3802.

Seelamantula, C. S. and Mulleti, S. 2014. Super-resolution reconstruction in frequency-domain optical-coherence tomography using the finite-rate-of-innovation principle, *IEEE Trans. Signal Process.*, Vol. 62, no. 19: 5020-5029.

Bruce, P. 2017. Essential process control for chemical engineers: bookboon.com ISBN 978-87-403-1655-1.

Dragotti, P.L., Vetterli, M. and Blu, T. 2007. Sampling moments and reconstructing signals of finite rate of innovation: Shannon meets Strang-Fix, *IEEE Trans. Signal Process.*, Vol. 55, no. 5: 1741-1757.

Seelamantula, C.S., and Unser, M. 2008. A generalized sampling method for finite-rate-of-innovation-signal reconstruction, *IEEE Signal Process. Lett.*: 813-816.

Olkkonen, H. and Olkkonen, J.T. 2008. Measurement and reconstruction of impulse train by parallel exponential filters, *IEEE Signal Process. Lett.*, Vol. 15: 241-244.

Derek, A. 2014. An introduction to nonlinearity in control systems: bookboon.com ISBN 978-87-7681-790-9.

Walters, G. and Shen, X. 2003. Sampling with prolate spheroidal functions, *Sampling Theory in Signal and Image Process.*, Vol. 2, pp. 25-52.

Şenay, S., Chaparro, L.F. and Akan, A. 2008. Sampling and reconstruction of non-bandlimited signals using Slepian functions, *Proc. European Signal Process. Conf. (EUSIPCO).*

Gang, F. and Rogelio, L. 1999. Adaptive control systems. Butterworth-Heinemann Linacre House, Jordan Hill, Oxford OX2 8DP.

Vaidyanathan, P.P. and Vrcelj, B. 2001. On sampling theory of non-bandlimited signals, *Proc. 4th Int. Conf. on Sampling Theory and Applications.*

Ayush B., Felix, K. and Ramesh, R. 2020. On Unlimited Sampling and Reconstruction, IEEE TRANS. ON SIG. PROC. (10.1109/TSP.2020.3041955)

Shankar, S. and Marc, B. 1989. Adaptive control: stability, convergence, and robustness: Prentice Hall information and system sciences series.

Kwonhue, C. and Huaping, L. 2016. Sampling and reconstruction, in problem-based learning in communication systems using MatLab and Simulink, IEEE: 66-77, doi: 10.1002/9781119060239.ch7.

CHAPTER TWO

THE Z-TRANSFORM OF SYSTEMS

2.1. Introduction

The z-transform is the preferred tool for the analysis and synthesis of linear, stationary, sampled or discrete in time systems. It may be seen as a discrete analogue of the Laplace transform. This type of transform makes it possible to easily describe the discrete in time signals and the response of invariant linear systems subjected to various inputs. It is used to simplify discrete in time systems, e.g., digital signal processing, digital filter design, etc.

2.2. The z-transform

2.2.1 Relation between Laplace transform and z-transform

The Laplace transform is given by the following equation:

$$L\left[x_s(t)\right]=\int_{-\infty}^{+\infty} x_s(t)e^{-st}dt$$

$\sum_{k=-\infty}^{+\infty} x(k)\delta(t-kT_s)$ is used instead of $x_s(t)$. The previous equation becomes:

$$L\left[x_s(t)\right]=\sum_{k=-\infty}^{+\infty} x(k)e^{-skT_s}$$

The z-transform of a discrete in time signal $x(k)$ is defined as follows

$$X(z)=\sum_{k=-\infty}^{+\infty} x(k)z^{-k} \qquad \text{or} \qquad X(z)=Z\left[x(k)\right]$$

where $z=e^{sT_s}$ is a complex variable ($z\in C$).

The causal discrete in time signal is defined as

$$x(k) = 0 \text{ for all } k < 0$$

so the uni-lateral z-transformation or the z-transform of a discrete in time signal $x(k)$ is defined as follows

$$X(z) = \sum_{k=0}^{+\infty} x(k) z^{-k}$$

Example:

The z-transform of the following discrete in time signal

$$x(k) = \frac{1}{3^k} u(k)$$

is given as

$$X(z) = \frac{3z}{3z-1}$$

or, using MatLab's function, the z-transform it is given as

```
>> syms z k
ztrans(1/3^k)
ans = 3*z/(3*z-1)
```

2.2.2 The region of convergence

The region of convergence (ROC) of $X(z)$ is the set of all values of z for which $X(z)$ attains a finite value. Mathematically speaking, the ROC is the set of values $z \in C$ for which the sequence $x(k)z^{-k}$ is, absolutely, summable, i.e.,

$$ROC = \left\{ z \in C : \left| X(z) \right| = \sum_{k=0}^{+\infty} \left| x(k) z^{-k} \right| < +\infty \right\}$$

In practice, it is not always necessary to specify the region of convergence of a certain z-transform, provided it is known that the series converges in some region.

The ROC is always an annulus, i.e.,

$$\left\{ r_2 \le |z| \le r_1 \right\}$$

where r_2 can be zero and r_1 can be $+\infty$.

Of according to the causal signal, the ROC is the exterior of a circle of some radius r_2, then it is defined by the unit circle in a complex z-plane.

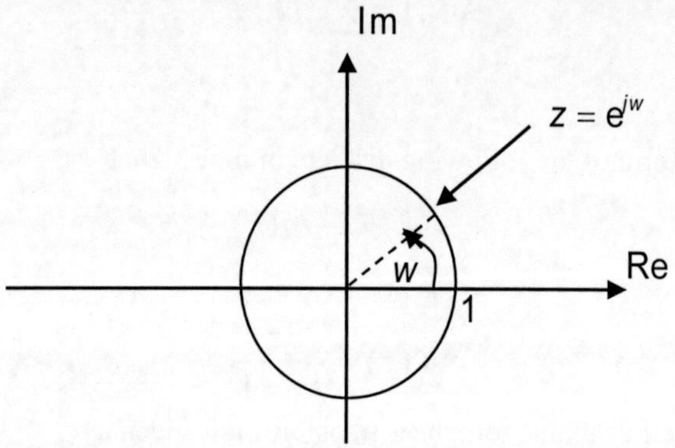

Fig. 2.1. The ROC of the causal signal.

The ROC is an important concept in many aspects: it allows the unique inversion of the z-transform and provides convenient characterizations of the causality and the stability properties of a signal or a system.

2.3 The z-transform by residual method

Let us consider a continuous in time signal given by the following equation

$$X(s) = \frac{N(s)}{D(s)}$$

where $N(s)$ and $D(s)$ are respectively the numerator and the denominator such that $\deg N(s) \leq \deg D(s)$.

If the continuous in time signal $X(s)$ has single poles, then the z-transform is

$$X(z) = \sum_{i=1}^{n} \left[\frac{N(s)}{D'(s)} \frac{1}{1 - e^{sT_s} z^{-1}} \right]_{s=p_i}$$

where p_i are simples poles of $X(s)$ and n is the number of poles.

Example: Let us consider a continuous-time signal $X(s)$ given as follows

$$H(s) = \frac{1}{s(s+2)}$$

We have two ($n=2$) simple poles of $X(s)$ which are $p_1 = 0$ and $p_2 = -2$.

$$X(z) = \sum_{i=1}^{2} \left[\frac{N(s)}{D'(s)} \frac{1}{1 - e^{sT_s} z^{-1}} \right]_{s=p_i}$$

$$= \frac{N(p_1)}{D'(p_1)} \frac{1}{1 - e^{p_1 T_s} z^{-1}} + \frac{N(p_2)}{D'(p_2)} \frac{1}{1 - e^{p_2 T_s} z^{-1}}$$

with:

$$N(s) = 1,$$
$$D(s) = s(s+2),$$
$$D'(s) = 2s + 2.$$

so

$$X(z) = \sum_{i=1}^{2} \left[\frac{N(s)}{D'(s)} \frac{1}{1 - e^{sT_s} z^{-1}} \right]_{s=p_i}$$

$$= \frac{1}{2} \left(\frac{1}{1 - z^{-1}} - \frac{1}{1 - e^{-2T_s} z^{-1}} \right)$$

$$= \frac{1}{2} \left(\frac{z}{z-1} - \frac{z}{z - e^{-2T_s}} \right)$$

Finally, we find

$$X(z) = \frac{1}{2} \frac{z(1 - e^{-2T_s})}{(z-1)(z - e^{-2T_s})}$$

If the continuous in time signal $X(s)$ has multiple poles of order m then the z-transform is

$$X(z)=\sum_{i=1}^{n}\frac{1}{(m-1)!}\left[\frac{d^{m-1}}{ds^{m-1}}\frac{X_i(s)}{1-e^{sT_s}z^{-1}}\right]_{s=p_i}$$

where p_i are simple poles of $X(s)$; n is the number of poles; m is the order of multiplicity; and

$$X_i(s)=(s-p_i)^m X(s)$$

Example: Let us consider a continuous in time signal $X(s)$ given as follows

$$X(s)=\frac{1}{s^2}$$

We have one simple pole ($n=1$) of $X(s)$ which is $p_1=0$ and the order of multiplicity is two ($m=2$).

$$X_1(s)=(s-p_1)^m X(s)$$

$$=s^2\frac{1}{s^2}$$

$$=1$$

So we get:

$$X(z)=\sum_{i=1}^{1}\frac{1}{(2-1)!}\left[\frac{d^{2-1}}{ds^{2-1}}\frac{1}{1-e^{sT_s}z^{-1}}\right]_{s=p_i}$$

$$=\frac{T_s z^{-1}e^{p_iT_s}}{(1-e^{p_iT_s}z^{-1})^2}$$

$$=\frac{T_s z^{-1}}{(1-z^{-1})^2}$$

$$=\frac{T_s z}{(z-1)^2}$$

2.4. Commonly used z-transform functions

2.4.1. Unit impulse function

Consider a discrete in time unit impulse function $\delta(k)$, defined by:

$$\delta(k)=\begin{cases}1 & if \quad k=0\\ 0 & if \quad k\neq 0\end{cases}$$

The z-transform of the discrete-time unit impulse function $\delta(k)$, is

$$Z\left\{\delta(k)\right\}=1+0z^{-1}+0z^{-2}+0z^{-3}+\ldots=1$$

2.4.2. Unit step function

Consider a discrete in time unit step function $u(k)$, defined by:

$$u(k)=\begin{cases}1 & if \quad k\geq 0\\ 0 & if \quad k<0\end{cases}$$

The z-transform of the discrete in time unit step function $u(k)$, is

$$Z\left\{u(k)\right\}=1+z^{-1}+z^{-2}+\ldots=\frac{1}{1-z^{-1}} \quad \text{if } \left|z^{-1}\right|<1$$

2.4.3. Unit ramp function

Consider a discrete in time unit ramp function $r(k)$, defined by:

$$r(k)=\begin{cases}T_s k & if \quad k\geq 0\\ 0 & if \quad k<0\end{cases}$$

The z-transform of the discrete in time unit ramp function $r(k)$, is

$$Z\left\{r(k)\right\}=\frac{T_s z^{-1}}{\left(1-z^{-1}\right)^2} \quad \text{if } \left|z^{-1}\right|<1$$

where T_s is a sampling period.

2.5. Properties of the z-transform

We focus now on listing the different properties of the z-transform. We start with the first one.

Linearity: if $X_1(z) = Z\{x_1(k)\}$ and $X_2(z) = Z\{x_2(k)\}$, then

$$\begin{aligned} Z\{a_1x_1(k)+a_2x_2(k)\} &= a_1Z\{x_1(k)\}+a_2Z\{x_2(k)\} \\ &= a_1X_1(z)+a_2X_2(z) \end{aligned}$$

with (a_1 and $a_2 \in \mathbf{R}$).

The proof of the linearity property is straightforward using obvious properties of the sum operation. By the z-transform definition:

$$\begin{aligned} Z\{a_1x_1(k)+a_2x_2(k)\} &= \sum_{k=0}^{+\infty}\left(a_1x_1(k)+a_2x_2(k)\right)z^{-k} \\ &= \sum_{k=0}^{+\infty}\left(a_1x_1(k)z^{-k}+a_2x_2(k)z^{-k}\right) \\ &= a_1\sum_{k=0}^{+\infty}\left(x_1(k)z^{-k}\right)+a_2\sum_{k=0}^{+\infty}\left(x_2(k)z^{-k}\right) \\ &= a_1Z\{x_1(k)\}+a_2Z\{x_2(k)\} \\ &= a_1X_1(z)+a_2X_2(z) \end{aligned}$$

Right shift theorem: let $X(z) = Z\{x(k)\}$, $d \in N$ and $x(k) = 0$, for $k < 0$, then

$$Z\{x(k-d)\} = z^{-d}Z\{x(k)\} = z^{-d}X(z)$$

Proof:

$$Z\{x(k-d)\}=\sum_{k=0}^{+\infty}\big(x(k-d)\big)z^{-k}$$

$$=\sum_{k=0}^{+\infty}x(k-d)z^{-k+d-d}$$

$$=z^{-d}\sum_{k=0}^{+\infty}x(k-d)z^{-(k-d)}$$

$$=z^{-d}\sum_{m=0}^{+\infty}x(m)z^{-m}$$

$$=z^{-d}Z\{x(k)\}$$

$$=z^{-d}X(z)$$

Left shift theorem: let $X(z)=Z\{x(k)\}$ and $d\in N$, then

$$Z\{x(k+d)\}=z^{d}\left(X(z)-\sum_{i=0}^{d-1}x(i)z^{-i}\right)$$

Proof:

$$Z\{x(k+d)\}=\sum_{k=0}^{+\infty}\big(x(k+d)\big)z^{-k}$$

let us consider $l=k+d$, then $k=l-d$. We find

$$Z\{x(k+d)\}=\sum_{l=d}^{+\infty}x(l)z^{-(l-d)}$$

$$=\sum_{l=d}^{+\infty}x(l)z^{-l}z^{d}$$

$$=z^{d}\sum_{l=d}^{+\infty}x(l)z^{-l}$$

$$=z^{d}\left(x(d)z^{-d}+x(d+1)z^{-(d+1)}+x(d+2)z^{-(d+2)}+...+x(d+n)z^{-(d+n)}\right)$$

$$Z\{x(k+d)\} = z^{d}\begin{pmatrix} x(d)z^{-d} + x(d+1)z^{-(d+1)} + ... + x(d+n)z^{-(d+n)} \\ +x(0)z^{-0} + x(1)z^{-1} + x(2)z^{-2} + ... + x(d-1)z^{-(d-1)} \\ -x(0)z^{-0} - x(1)z^{-1} - x(2)z^{-2} - ... - x(d-1)z^{-(d-1)} \end{pmatrix}$$

$$Z\{x(k+d)\} = z^{d}\left(X(z) - x(0)z^{-0} - x(1)z^{-1} - x(2)z^{-2} - ... - x(d-1)z^{-(d-1)}\right)$$

$$Z\{x(k+d)\} = z^{d}\left(X(z) - \sum_{i=0}^{d-1} x(i)z^{-i}\right)$$

Multiplication by k : let $X(z) = Z\{x(k)\}$, then

$$Z\{kx(k)\} = -z\frac{d[X(z)]}{dz}$$

Proof:

$$Z\{kx(k)\} = \sum_{k=0}^{+\infty}(kx(k))z^{-k}$$

$$=-\sum_{k=0}^{+\infty} x(k)(-k)z^{-k-1+1}$$

$$= -z\frac{d}{dz}\left[\sum_{k=0}^{+\infty} x(k)z^{-k}\right]$$

$$= -z\frac{d}{dz}[X(z)]$$

Multiplication by a^{k} : let $X(z) = Z\{x(k)\}$ and $a \in C$, then

$$Z\{a^{k}x(k)\} = X\left(\frac{z}{a}\right); \qquad a \neq 0$$

Proof:

$$Z\left\{a^k x(k)\right\} = \sum_{k=0}^{+\infty}\left(a^k x(k)\right) z^{-k}$$

$$= \sum_{k=0}^{+\infty} x(k)\left(a^{-1}z\right)^{-k}$$

$$= \sum_{k=0}^{+\infty} x(k)\left(\frac{z}{a}\right)^{-k}$$

$$= X\left(\frac{z}{a}\right)$$

In a similar way, we get: $Z\left\{a^{-k} x(k)\right\} = X(az)$.

Initial value theorem: if $x(k)$ is zero for $k < 0$, i.e., $x(k)$ is causal, then

$$x(0) = \lim_{z \to +\infty} X(z)$$

$$x(1) = \lim_{z \to +\infty} z\left[X(z) - x(0)\right]$$

$$x(2) = \lim_{z \to +\infty} z^2\left[X(z) - x(0) - x(1)z^{-1}\right]$$

$$\vdots$$

Proof:
By the z-transform definition:

$$X(z) = \sum_{k=0}^{+\infty} x(k)z^{-k} = x(0) + x(1)z^{-1} + x(2)z^{-2} + \ldots$$

Taking the limit for the modulus of 2 of z when tending it to infinity, we obtain the result specified by the initial value theorem.

$$x(0) = \lim_{z \to +\infty} X(z) = \lim_{z \to +\infty}\left[x(0) + x(1)z^{-1} + x(2)z^{-2} + \ldots\right] = x(0) + 0 + 0 + ..$$

By the way

$$X(z) - x(0) = x(1)z^{-1} + x(2)z^{-2} + x(3)z^{-3} + \ldots$$

and

$$z\left[X(z) - x(0)\right] = x(1) + x(2)z^{-1} + x(3)z^{-2} + \ldots$$

then

$$\lim_{z\to+\infty} z\left[X(z)-x(0)\right]=\lim_{z\to+\infty}\left[x(1)+x(2)z^{-1}+x(3)z^{-2}+\ldots\right]=x(1)+0+0+..$$

Using the same way, we get that:

$$x(2)=\lim_{z\to+\infty} z^{2}\left[X(z)-x(0)-x(1)z^{-1}\right]$$

Final value theorem: if $Z\left[x(k)\right]=X(z)$, then

$$\begin{aligned} x(+\infty) &= \lim_{k\to+\infty} x(k) \\ &= \lim_{z\to1}(z-1)X(z) \end{aligned}$$

Proof:

$$\begin{aligned} \sum_{k=0}^{+\infty}\left(x(k+1)-x(k)\right)z^{-k} &= \sum_{k=0}^{+\infty} x(k+1)z^{-k}-\sum_{k=0}^{+\infty} x(k)z^{-k} \\ &= \sum_{m=1}^{+\infty} x(m)z^{-(m-1)}-X(z) \\ &= z\left(\sum_{m=0}^{+\infty} x(m)z^{-m}-x(0)\right)-X(z) \end{aligned}$$

So

$$\begin{aligned} \sum_{k=0}^{+\infty}\left(x(k+1)-x(k)\right)z^{-k} &= z\left(X(z)-x(0)\right)-X(z) \\ &= X(z)\left(z-1\right)-zx(0) \end{aligned}$$

When z converges to 1, the equation gives:

$$\lim_{z\to1}\left(\sum_{k=0}^{+\infty}\left(x(k+1)-x(k)\right)z^{-k}\right)=\lim_{z\to1}\left(X(z)\left(z-1\right)\right)-x(0)$$

or:

$$x(+\infty)-x(0)=\lim_{z\to1}\left(X(z)\left(z-1\right)\right)-x(0).$$

Then:

$$x(+\infty)=\lim_{z\to1}\left(X(z)\left(z-1\right)\right)$$

Convolution: let $x_1(k)$ and $x_2(k)$ be two sequences, we call $x(k)$ a convolution product defined by:

$$x(k)=\sum_{j=0}^{+\infty} x_1(j)x_2(k-j)$$
$$=\sum_{j=0}^{+\infty} x_1(k-j)x_2(j)$$
$$=x_1(k)\otimes x_2(k)$$

The z-transform of the convolution product $x(k)$ is a scalar product:

$$Z\{x_1(k)\otimes x_2(k)\}=X_1(z)X_2(z)$$

Proof:

$$X(z)=\sum_{k=0}^{+\infty} x(k)z^{-k}$$
$$=\sum_{k=0}^{+\infty}\left[\sum_{j=0}^{+\infty} x_1(j)x_2(k-j)\right]z^{-k}$$
$$=\sum_{j=0}^{+\infty} x_1(j)\left[\sum_{k=0}^{+\infty} x_2(k-j)z^{-k}\right]$$
$$=\sum_{j=0}^{+\infty} x_1(j)z^{-j}X_2(z)$$
$$=X_1(z)X_2(z)$$

with $X(z)=Z\{x(k)\}$, $X_1(z)=Z\{x_1(k)\}$ and $X_2(z)=Z\{x_2(k)\}$.

2.6. Examples of z-transform

Example 1

Find the z-transform and the radius of convergence for each discrete in time signal

$$x_1(k) = \begin{cases} 1 & if \quad k = 0 \\ 0 & if \quad k \neq 0 \end{cases}$$

$$x_2(k) = \begin{cases} 1 & if \quad k \geq 0 \\ 0 & if \quad k < 0 \end{cases}$$

$$x_3(k) = kx_2(k) \quad k \geq 0$$

$$x_4(k) = a^k x_2(k), \; k \geq 0, \; a \in C^*$$

$$x_5(k) = ke^{-ak} x_2(k)$$

$$x_6(k) = \sin(bk) x_2(k), \; k \geq 0.$$

Solution

1- The z-transform of $x_1(k)$ is:

$$X_1(z) = Z\{x_1(k)\} = \sum_{k=0}^{+\infty} x_1(k) z^{-k} = 1$$

$ROC = C$ (entire z-plane or $|z| > 0$).

2- The z-transform of $x_2(k)$ is:

$$X_2(z) = Z\{x_2(k)\} = \sum_{k=0}^{+\infty} x_2(k) z^{-k} = \frac{z}{z-1}$$

The ROC is $|z| > 1$.

3- The z-transform of $x_3(k)$ is:

$$\begin{aligned} X_3(z) &= Z\{x_3(k)\} \\ &= Z\{kx_2(k)\} \\ &= -z \frac{dX_2(z)}{dz} \\ &= \frac{z}{(z-1)^2} \end{aligned}$$

The ROC is $|z| > 1$.

4- The z-transform of $x_4(k)$ is:

$$X_4(z) = Z\{x_4(k)\}$$

$$= \sum_{k=0}^{+\infty} x_4(k) z^{-k}$$

$$= \sum_{k=0}^{+\infty} a^k x_2(k) z^{-k}$$

$$= \sum_{k=0}^{+\infty} a^k z^{-k}$$

$$= \sum_{k=0}^{+\infty} (\frac{a}{z})^k$$

let $v = \frac{a}{z}$, then

$$X_4(z) = \sum_{k=0}^{+\infty} (v)^k$$

$$= 1 + v + v^2 + v^3 + ... + v^n$$

$$X_4(z) = \frac{1 - v^{n+1}}{1 - v} = \frac{1}{1 - v} \quad \text{if } n \to +\infty \text{ and } |v| < 1$$

$$= \frac{z}{z - a}$$

The ROC is $|z| > |a|$.

5- The z-transform of $x_5(k)$ is:

$$X_5(z) = Z\left\{ke^{-ak} x_2(k)\right\}$$
$$= Z\left\{e^{-ak} x_3(k)\right\}$$
$$= X_3(\frac{z}{a})$$
$$= \frac{\frac{z}{a}}{\left(\frac{z}{a}-1\right)^2}$$
$$= \frac{az}{\left(z-a\right)^2}$$

The ROC is $|z|>|a|$.

6- The z-transform of $x_6(k)$ is:

$$X_6(z) = Z\left\{x_6(k)\right\}$$
$$= Z\left\{\sin(bk)\right\}$$
$$= Z\left\{\frac{1}{2i}(e^{ibk} - e^{-ibk})\right\}$$
$$= \frac{1}{2i}\left(Z\left\{\left(e^{ib}\right)^k x_2(k)\right\} - Z\left\{\left(e^{-ib}\right)^k x_2(k)\right\}\right)$$
$$= \frac{1}{2i}\left(\frac{z}{z-e^{ib}} - \frac{z}{z-e^{-ib}}\right)$$
$$= \frac{z\sin(b)}{z^2 - 2z\cos(b) + 1}$$

The ROC is $|z|>1$.

Example 2: Let us consider the following discrete-time signal.

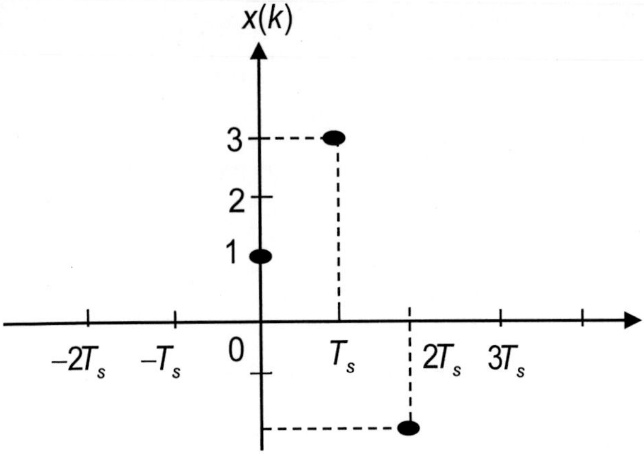

Fig. 2.2. A discrete-time signal

1- Find the z-transform of the discrete in time signal $x(k)$.
2- Find the z-transform of the discrete in time signal $x(k+1)$ and $x(k+2)$.

Solution

1- The z-transform of the discrete in time signal is:

$$X(z)=Z\{x(k)\}=\sum_{k=0}^{+\infty}x(k)z^{-k}=x(0)z^{-0}+x(T_s)z^{-1}+x(2T_s)z^{-2}=1+3z^{-1}-2z^{-2}$$

2- The z-transform of the discrete in time signal $x(k+1)$ is:

$$X_1(z)=Z\{x(k+1)\}=z^1(X(z)-\sum_{i=0}^{1-1}x(i)z^{-i})$$

$$=z(1+3z^{-1}-2z^{-2}-x(0))=z(1+3z^{-1}-2z^{-2}-1)=3-2z^{-1}$$

The z-transform of the discrete in time signal is $x(k+2)$:

$$X_2(z)=Z\{x(k+2)\}=z^2(X(z)-\sum_{i=0}^{2-1}x(i)z^{-i})$$

$$=z^2(1+3z^{-1}-2z^{-2}-x(0)z^{-0}-x(1)z^{-1})=z^2(1+3z^{-1}-2z^{-2}-1-3z^{-1})=-2$$

Example 3

1 - Let the discrete in time signal be defined by $X_1(z)=\dfrac{z}{z-a}$ and apply the theorem of the initial value.

2 - Let the discrete in time signal be defined by $X_2(z)=\dfrac{0.5z}{(z-1)(z-0.5)}$ and then apply the theorem of the final value.

Solution

1 - We apply the theorem of the initial value:

$$x_1(0)=\lim_{z\to+\infty} X_1(z)=\lim_{z\to+\infty}\frac{z}{z-a}=1$$

2 - We apply the theorem of the final value:

$$x_2(+\infty)=\lim_{z\to1}\,(z-1)X_2(z)=\lim_{z\to1}\,(\text{z-1})\frac{0.5z}{(z-1)(z-0.5)}=1$$

Example 4

Let the discrete in time signal be defined by $x_1(k)=\{1;3;-2\}$ and $x_2(k)=\{1.5;-1;0.5\}$, then calculate the convolution product $x(k)=x_1(k)\otimes x_2(k)$.

Solution

Let the discrete in time signal be defined by $x_1(k)=\{1;3;-2;0;0\}$ and $x_2(k)=\{1.5;-1;0.5;0;0\}$,

$$\begin{aligned}X(z)&=X_1(z).X_2(z)\\&=\left(1+3z^{-1}-2z^{-2}\right)\left(1.5-z^{-1}+0.5z^{-2}\right)\\&=1.5-z^{-1}+0.5z^{-2}+4.5z^{-1}-3z^{-2}+1.5z^{-3}-3z^{-2}+2z^{-3}-z^{-4}\\&=1.5+3.5z^{-1}-5.5z^{-2}+3.5z^{-3}-z^{-4}\end{aligned}$$

hence the product of convolution is:

$$x(k) = x_1(k) \otimes x_2(k)$$
$$= \{1.5; 3.5; -5.5; 3.5; -1\}$$

2.7 The inverse z-transform

The z-transform is a mapping from a sequence $\{x(k)\}$ to a complex function $X(z)$. This mapping is useful only if it is invertible, i.e., from a given $X(z)$ it is possible to find, in a unique way, the sequence $\{x(k)\}$ such that $Z\{x(k)\} = X(z)$.

The sequence $\{x(k)\}$ is referred to as the inverse z-transform of $X(z)$, which is given as

$$\{x(k)\} = Z^{-1}[X(z)] \qquad \text{or} \qquad x(k) = Z^{-1}[X(z)]$$

The inverse z-transform of $X(z)$ can be computed several ways.

2.7.1. The inverse z-transform by inspection method

This method has basically become familiar with the z-transform pair tables.

Example: The inverse z-transform of the following function

$$X(z) = \frac{3z}{3z-1}$$

is

$$x(k) = Z^{-1}[X(z)] = \left(\frac{1}{3}\right)^k$$

Using the MatLab function, the inverse z-transform is

```
>> syms z k
iztrans(3*z/(3*z-1))
ans = (1/3)^n
```

This method is one of the simplest methods for finding the inverse z-transform and can be used for almost every type of expression given in fractional form. It is also called the power series expansion method.

If we calculate the result of the polynomial division according to the increasing terms of z^{-1}, we obtain

$$\begin{aligned} X(z) &= \frac{N(z)}{D(z)} \\ &= \sum_{k=0}^{+\infty} C(k) z^{-k} \\ &= C_0 + C_1 z^{-1} + C_2 z^{-2} + \ldots \\ &= x(0) + x(1) z^{-1} + x(2) z^{-2} + \ldots \end{aligned}$$

We identify the quotient with the equation

$$\begin{aligned} x(0) &= C_0 \\ x(1) &= C_1 \\ x(2) &= C_2 \\ &\vdots \end{aligned}$$

then by the uniqueness of the z-transform, the inverse z-transform is

$$x(k) = c_k$$

i.e., the signal sample values in the time-domain are the corresponding coefficients of the power series expansion.

Example: In the inverse z-transform of the following functions

$$X_1(z) = \frac{z}{z^2 - 3z + 2} \quad \text{and} \quad X_2(z) = \frac{4z^2 + 2z}{2z^2 - 3z + 1}$$

$X_1(z)$ could be given as follows

$$\begin{aligned} X_1(z) &= \frac{z}{z^2 - 3z + 2} \\ &= z^{-1} + 3z^{-2} + 7z^{-3} + \ldots. \\ &= \sum_{k=0}^{+\infty} \left(2^k - 1\right) z^{-k} \end{aligned}$$

So the inverse z-transform is

$$x_1(k) = 2^k - 1$$

$X_2(z)$ could be

$$X_2(z) = \frac{4z^2 + 2z}{2z^2 - 3z + 1}$$
$$= 2 + 4z^{-1} + 5z^{-2} + \ldots$$
$$= \sum_{k=0}^{+\infty} \left(6 - 2^{2-k}\right) z^{-k}$$

so the inverse z-transform is

$$x_2(k) = 6 - 4(\frac{1}{2})^k$$

Example: Find the inverse z-transform of $X(z)$

$$X(z) = \frac{1}{1 - 2z^{-1}}$$

Using the properties of the geometric series we get:

$$X(z) = \frac{1}{1 - 2z^{-1}}$$
$$= \sum_{k=0}^{+\infty} (2)^k z^{-k}$$
$$= 1 + 2z^{-1} + 2^2 z^{-2} + 2^3 z^{-3} + 2^4 z^{-4} + \ldots$$

Finally, the coefficients of the power series expansion are

$$x(k) = \{1, 2, 4, 8, \ldots\}$$

This method does not give the general term of the sequence $\{x(k)\}$. It calculates each term according to its rank in the expression. However, it is very practical and well suited to the analysis of the transient regimes of discrete or sampled systems.

2.7.3. The inverse z-transform by the partial-fraction expansion method

The general idea of this method is to find for the function $X(z)$ a development in simpler z-functions for which an inverse transform is known.

The method relies on the linearity of the z-transform and on the representation of the function $X(z)$ in a special form.

Step 1: Decompose X(z) **into proper form**

The rational z-transform $X(z)$ is given by:

$$X(z)=\frac{b_0 z^M+b_1 z^{M-1}+\ldots+b_M}{z^N+a_1 z^{N-1}+\ldots+a_N}=\frac{b_0+b_1 z^{-1}+\ldots+b_M z^{-M}}{1+a_1 z^{-1}+\ldots+a_N z^{-N}}$$

A rational function is called proper if $a_N \neq 0$ and $M \leq N$.

Step 2: Find simple elements of $\frac{X(z)}{z}$

This method consists on breaking down $\frac{X(z)}{z}$ into simple elements as follows

$$\frac{X(z)}{z}=\frac{r_0}{z}+\frac{r_1}{z-p_1}+\frac{r_2}{z-p_2}+\ldots+\frac{r_n}{z-p_n}$$

with $p_i \neq p_j$, $i \neq j$ and $p_i \neq 0$, for all i,

$$r_0 = X(0)$$

and

$$r_i=\lim_{z\to p_i}\left(z-p_i\right)\frac{X(z)}{z}.$$

then

$$X(z)=r_0+\frac{zr_1}{z-p_1}+\frac{zr_2}{z-p_2}+\ldots+\frac{zr_n}{z-p_n}$$

Step 3: Inverse z-transform

Using the following equation

$$Z\left[(p_i)^k\right]=\frac{z}{z-p_i}$$

and

$$Z\left[\delta(k)\right]=1 \text{ for } k=0$$

we get

$$x(k)=r_0\delta(k)+\sum_{i=1}^{n} r_i(p_i)^k$$

Note that, since $X(z)$ has real coefficients, complex poles appear in conjugate pairs, hence the corresponding residuals are also complex, and the sequence $\{x(k)\}$ has real valued terms.

Example: Let us consider the following z-transform function

$$\text{X(z)}=\frac{0.2z}{z^2-0.7z+0.12}$$

$X(z)$ has 2 simple poles at $z=0.3$ and $z=0.4$. The system is stable.

Now express $\frac{X(z)}{z}$ as

$$\frac{\text{X(z)}}{\text{z}}=\frac{0.2}{(z-0.3)(z-0.4)}$$

The partial fraction expansion of $\frac{X(z)}{z}$ is

$$\frac{\text{X(z)}}{\text{z}}=-\frac{2}{z-0.3}+\frac{2}{z-0.4}$$

The partial fraction expansion of $X(z)$ is

$$X(z)=-\frac{2z}{z-0.3}+\frac{2z}{z-0.4}$$

The inverse transform of the expression is

$$x(k)=2\left((0.4)^k-(0.3)^k\right)u(k)$$

Or one can use MatLab functions, and the z-transform function is

```
>> [r,p,k]=residue([0.2 0],[1 -0.7 0.12])
r =
    0.8000
   -0.6000
p =
    0.4000
    0.3000
k =
    []
```

2.7.4. The inverse z-transform by recurrence inversion method

Let $X(z)$ be a function given as follows

$$X(z)=\frac{N(z^{-1})}{D(z^{-1})}=\frac{c_0+c_1z^{-1}+c_2z^{-2}+\ldots+c_Mz^{-M}}{d_0+d_1z^{-1}+d_2z^{-2}+\ldots+d_Nz^{-N}}$$

The inverse z-transform is:

$$x(kT_s)=\frac{1}{d_0}\left[c_k-\sum_{i=0}^{k-1}x(iT_s)d_{k-i}\right]$$

This method does not give the general term of the sequence $\{x(k)\}$. It allows each term to be calculated according to its rank. However, it is very practical and well suited to the analysis of the transient regimes of discrete or sampled systems.

Example: Let us consider the following z-transform function

$$X(z)=\frac{0.2z}{z^2-0.7z+0.12}=\frac{0.2z^{-1}}{1-0.7z^{-1}+0.12z^{-2}}$$

$$X(z)=\frac{c_0+c_1z^{-1}}{d_0+d_1z^{-1}+d_2z^{-2}}$$

with $c_0=0$, $c_1=0.2$, $d_0=1$, $d_1=-0.7$ and $d_2=0.12$ from where the first four samples of $x(k)$ are:

$$x(0) = \frac{c_0}{d_0} = 0$$

$$x(1) = \frac{1}{d_0}\left[c_1 - x(0)d_0\right] = \frac{c_1}{d_0} = \frac{0.2}{1} = 0.2$$

$$x(2) = \frac{1}{d_0}\left[c_2 - x(0)d_2 - x(T_s)d_1\right] = (-0.2)\times(-0.7) = 0.14$$

$$x(3) = \frac{1}{d_0}\left[c_3 - x(0)d_3 - x(T_s)d_2 - x(2T_s)d_1\right] = (-0.2)\times 0.12 - 0.14\times(-0.7) = 0.122$$

2.7.5. The inverse z-transform by contour integration method

The z-transform is given as follows.

$$X(z) = \sum_{k=0}^{+\infty} x(k)z^{-k} = x(0) + x(1)z^{-1} + \ldots + x(k)z^{-k}$$

Its inverse z-transform is defined as

$$x(k) = Z^{-1}\left[X(z)\right]$$

$$= \frac{1}{2\pi j}\int_{\Gamma} z^{k-1}X(z)dz$$

where $\int_{\Gamma} z^{k-1}X(z)dz$ is the contour integral and Γ is any contour around the origin.

Due to complex functions properties, the above integral can be rewritten as a sum of residues as shown below:

$$x(k) = \sum\left[\text{Residues of } z^{k-1}X(z) \text{ at the poles of } X(z)\right]$$

Example: consider the following z-transform function

$$\mathrm{X(z)} = \frac{6z}{6z^2 - 5z + 1}$$

Divide $X(z)$ by z to expand the expression in partial fractions.

$$\frac{X(z)}{z} = \frac{6}{6z^2 - 5z + 1}$$

Substituting into the inversion integral gives

$$x(k)=\frac{1}{2\pi j}\left[\int_{\Gamma}\frac{6z}{(2z-1)(3z-1)}z^{k-1}dz\right]$$

$$=\frac{1}{2\pi j}\left[\int_{\Gamma}\left(\frac{6z^k}{z-\frac{1}{2}}-\frac{6z^k}{z-\frac{1}{3}}\right)dz\right]$$

The first integral can be evaluated as:

$$2\pi j\left[\frac{(z-\frac{1}{2})6z^k}{z-\frac{1}{2}}\right]_{z\to\frac{1}{2}}\Rightarrow 2\pi j6(\frac{1}{2})^k$$

The second integral can be evaluated as:

$$2\pi j\left[\frac{(z-\frac{1}{3})6z^k}{z-\frac{1}{3}}\right]_{z\to\frac{1}{3}}\Rightarrow 2\pi j6(\frac{1}{3})^k$$

The total integral gives the inverse z-transform:

$$x(k)=6\left[\left(\frac{1}{2}\right)^k-\left(\frac{1}{3}\right)^k\right]$$

This method is usually not straightforward since it involves the evaluation of a contour integral. In this course, we will not use the method of contour integrals to actually compute inverse z-transforms (such contour integrals are difficult to evaluate), therefore, we often use other techniques to obtain the inverse z-transform.

2.8. Examples of inverse z-transform

Find the first four samples of $x(k)$ of the z-transform

$$X(z)=\frac{2z}{z^2-1.3z+0.4}$$

2.8.1 The inverse z-transform by the inspection method

The inverse z-transform of the following function

$$X(z) = \frac{2z}{z^2 - 1.3z + 0.4}$$

is

$$x(k) = Z^{-1}\left[X(z)\right] = \frac{20}{3}\left[\left(\frac{4}{5}\right)^k - \left(\frac{1}{2}\right)^k\right]$$

The first four samples of $x(k)$ of the z-transform are: $x(0) = 0$, $x(1) = 2$, $x(2) = 2.6$, and $x(3) = 2.58$.

Using the MatLab function, the inverse z-transform is

```
>> syms z k
>> iztrans(2*z/(z^2-1.3*z+0.4))
ans =(20*(4/5)^k)/3 - (20*(1/2)^k)/3
```

2.8.2 The inverse z-transform by the direct division method

The inverse z-transform of the following function

$$X(z) = \frac{2z}{z^2 - 1.3z + 0.4}$$

is

$2z$	$z^2 - 1.3z + 0.4$
	$2z^{-1} + 2.6z^{-2} + 2.58z^{-3} + \ldots$

So the first four samples of $x(k)$ of the z-transform are: $x(0) = 0$, $x(1) = 2$, $x(2) = 2.6$, and $x(3) = 2.58$.

We can use MatLab which gives the inverse z-transform of the function:

```
>> n=4
>> a=[1 -1.3 0.4];
>> b=[0 2 0];
>> b=[b zeros(1,n-1)];
>> [x,r]=deconv(b,a);
>> disp(x)
     0    2.0000    2.6000    2.5800
```

2.8.3 The inverse z-transform by the partial-fraction expansion method

To find the first four samples of $x(k)$ using the partial-fraction expansion method, let us consider the following expression:

$$X(z)=\frac{2z}{z^2-1.3z+0.4}=\frac{2z}{(z-0.8)(z-0.5)}$$

We will break $\frac{\mathrm{X(z)}}{\mathrm{z}}$ into simple elements:

$$\frac{\mathrm{X(z)}}{\mathrm{z}}=\frac{2}{(\mathrm{z}-0.8)(\mathrm{z}-0.5)}=\frac{\mathrm{A}_1}{(\mathrm{z}-0.8)}+\frac{\mathrm{A}_2}{(\mathrm{z}-0.5)}$$

with:

$$\mathrm{A}_1=\frac{2}{(\mathrm{z}-0.5)}\Big\|_{\mathrm{z}=0,8}=\frac{20}{3}$$

$$\mathrm{A}_2=\frac{2}{(\mathrm{z}-0.8)}\Big\|_{\mathrm{z}=0,5}=-\frac{20}{3}$$

Then we get

$$\frac{\mathrm{X(z)}}{\mathrm{z}}=\frac{\frac{20}{3}}{\mathrm{z}-0.8}-\frac{\frac{20}{3}}{\mathrm{z}-0.5}$$

Finally we obtain

$$X(z) = \frac{20}{3}\left(\frac{z}{z-0.8} - \frac{z}{z-0.5}\right)$$

The inverse z-transform of $X(z)$ is:

$$x(k) = \frac{20}{3}\left[(0.8)^k - (0.5)^k\right]$$

Hence, the first four samples of $x(k)$ are:

$$x(0) = \frac{20}{3} \times 0 = 0,$$

$$x(1) = \frac{20}{3}(0.8 - 0.5) = 2,$$

$$x(2) = \frac{20}{3}((0.8)^2 - (0.5)^2) = 2.6$$

and

$$x(3) = \frac{20}{3}((0.8)^3 - (0.5)^3) = 2.58$$

2.8.4 The inverse z-transform by recurrence inversion method

In order to find the first four samples of $x(k)$ using the recurrence inversion method, let us consider the following expression:

$$X(z) = \frac{2z}{0.4 - 1.3z + z^2}$$

$$= \frac{c_0 + c_1 z^{-1}}{d_0 + d_1 z^{-1} + d_2 z^{-2}}$$

with $c_0 = 0$, $c_1 = 2$, $d_0 = 1$, $d_1 = -1.3$ and $d_2 = 0.4$, from where the first four samples of $x(k)$ are:

$$x(0) = \frac{c_0}{d_0} = 0$$

$$x(1) = \frac{1}{d_0}\left[c_1 - x(0)d_0\right] = \frac{c_1}{d_0} = \frac{2}{1} = 2$$

$$x(2)=\frac{1}{d_0}\left[c_2-x(0)d_2-x(T_s)d_1\right]=-2\times(-1.3)=2.6$$

$$x(3)=\frac{1}{d_0}\left[c_3-x(0)d_3-x(T_s)d_2-x(2T_s)d_1\right]=0-0-2\times0.4-2.6\times(-1.3)=2.58$$

2.8.5 The inverse z-transform by contour integration method

To find the first four samples of $x(k)$ using the contour integration method, let us consider the following expression:

$$X(z)=\frac{2z}{z^2-1.3z+0.4}$$

divide $X(z)$ by z to expand in partial fractions.

$$\frac{X(z)}{z}=\frac{2}{z^2-1.3z+0.4}$$

Substituting into the inversion integral gives

$$x(k)=\frac{1}{2\pi j}\left[\int_\Gamma\frac{2z}{(z-0.8)(z-0.5)}z^{k-1}dz\right]$$

$$=\frac{1}{2\pi j}\frac{20}{3}\left[\int_\Gamma\left(\frac{z^k}{z-0.8}-\frac{z^k}{z-0.5}\right)dz\right]$$

The first integral can be evaluated as

$$2\pi j\left[\frac{(z-0.8)z^k}{z-0.8}\right]_{z\to0.8}\Rightarrow2\pi j(0.8)^k$$

The second integral can be calculated as:

$$2\pi j\left[\frac{(z-0.5)z^k}{z-0.5}\right]_{z\to0.5}\Rightarrow2\pi j(0.5)^k$$

The total integral gives the inverse z-transform:

$$x(k)=\frac{20}{3}\left[(0.8)^k-(0.5)^k\right]$$

Table 2.1 summarizes the z-transforms of the most used functions in signal processing. T_s is the sampling period of the transformed signal in which $t = kT_s$ has been set.

Table 2.1. The z-transforms of the most used functions in signal processing

$x(k)\ \forall k \geq 0$	$X(z)$	**ROC**
$\delta(k)$	1	$\lvert z\rvert > 0$
$u(k)$	$\dfrac{z}{z-1}$	$\lvert z\rvert > 1$
$ku(k)$	$\dfrac{T_s z}{(z-1)^2}$	$\lvert z\rvert > 1$
$a^k u(k)$	$\dfrac{z}{z-a}$	$\lvert z\rvert > \lvert a\rvert$
$e^{-ak} u(k)$	$\dfrac{z}{z-e^{-aT_s}}$	$\lvert z\rvert > \lvert e^{-a}\rvert$
$ke^{-ak} u(k)$	$\dfrac{T_s az}{(z-1)^2}$	$\lvert z\rvert > \lvert e^{-a}\rvert$

2.9 Conclusions

In this chapter, some definitions of the z-transform and the inverse z-transform approach are detailed. Indeed, the inspection method, the direct division method, the partial-fraction expansion method, the recurrence inversion method and the contour integration method are demonstrated.

2.10 References

Zhengtao, D. 2013. Nonlinear and adaptive control systems: The Institution of Engineering and Technology.

Alan, V.O. Alan, S.W. and Hamind, N.S. 2005. Signals and systems, Second Edition, Pearson Education, 8th Indian Reprint.

Gang, F. and Rogelio, L. 1999. Adaptive control systems. Butterworth-Heinemann Linacre House, Jordan Hill, Oxford OX2 8DP.

John, G.P. and Dimitris, G.M. 2007. Digital signal processing: principles, algorithms, and applications, 4th edition.

Derek, A. 2014. An introduction to nonlinearity in control systems: bookboon.com ISBN 978-87-7681-790-9.

Andrew, C.S., John, R.B., Michael, M.D. 1996. Computer explorations in signals and systems using MatLab, Prentice Hall Signal Processing Series Paperback – January 1.

Shankar, S. and Marc, B. 1989. Adaptive control: stability, convergence, and robustness: Prentice Hall information and system sciences series.

Ramakrishna, P.R. 2008. Signals and systems, Tata McGraw-Hill.

Roberts, M.J. 2011. Signals and systems, analysis using transform methods and MatLab, Second edition, McGraw-Hill Education.

Bruce, P. 2017. Essential process control for chemical engineers: bookboon.com ISBN 978-87-403-1655-1.

Sunetra, S.A., Mona, V.D. 2015. Relation of z-transform and Laplace transform in discrete time signal, International Research Journal of Engineering & Technology, Vol. 2, Issue 2.

Tarun, K.R. 2011. Signals and systems, Oxford University Press.

Arild, W. 2013. Discrete dynamical systems with an introduction to discrete optimization problems: bookboon.com ISBN 978-87-403-0327-8.

Mitra, S.K. 2006. Digital signal processing. McGraw-Hill, New York.

Hsu, H. 2011. Signals and systems, Schaum's outlines, 2nd edn. McGraw-Hill, New York.

Kailath, T. 1980. Linear systems. Prentice-Hall, Englewood Cliffs.

Derek, A. 2015. Control engineering problems with solutions: bookboon.com ISBN 978-87-403-0374-2.

Ezeagwu Christopher, O.E.D., Mmuo Grace, C., Onu, C. 2019. Analysis and relevance of z-transform in discrete time systems. International Journal of Scientific & Engineering Research Vol. 10, Issue 11: 1163-1168.

CHAPTER THREE

THE DISCRETE-TIME TRANSFER FUNCTION

3.1. Introduction

In this chapter, we will discuss the transfer function of a linear time invariant (LTI) discrete-time system which can be used to represent the system in a simulator or in computer tools for analysis and design (such as SIMULINK and MatLab).

3.2. Transfer function from impulse response

3.2.1. Impulse function

An impulse function is also known as a Dirac delta function. A system transforms an input signal, $u(k)$, into an output signal, $y(k)$.

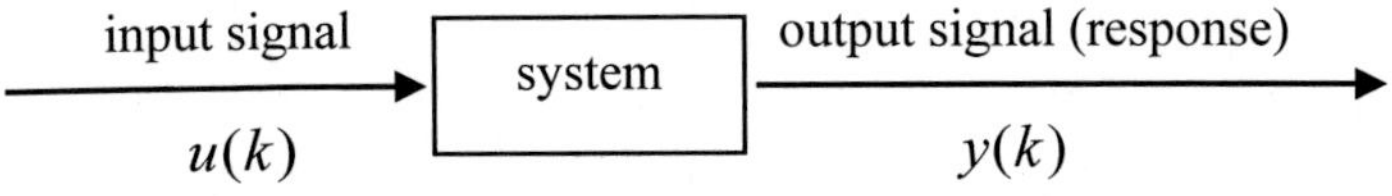

Fig. 3.1. The relation between input and output of system

If an input signal $u(k)=\delta(k)$, the response of the system is called an impulse response and $y(k)=h(k)$.

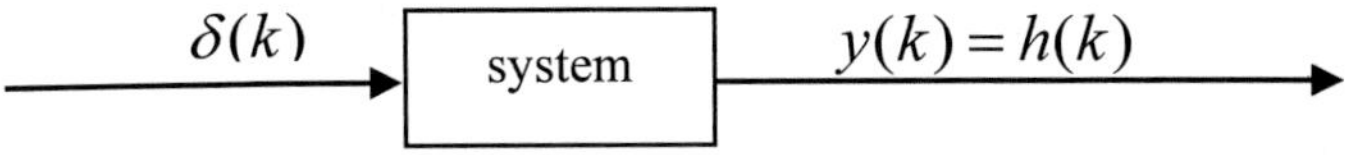

Fig. 3.2. The impulse response

If the input signal is a step function, then the response of the system is called a step response.

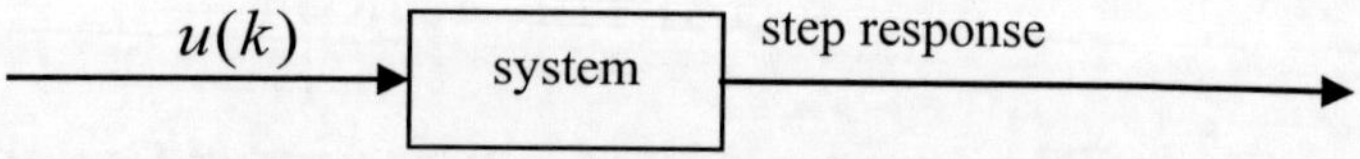

Fig. 3.3. The step response

The impulse response is important and very useful. If we know the response of a system to an impulse function, then we can know the response of the same system to any kind of input (step signal, ramp signal, and random signal).

3.2.2. Impulse response of system

The response of a given linear time invariant system to the input $u(k)$ is the output $y(k)$.

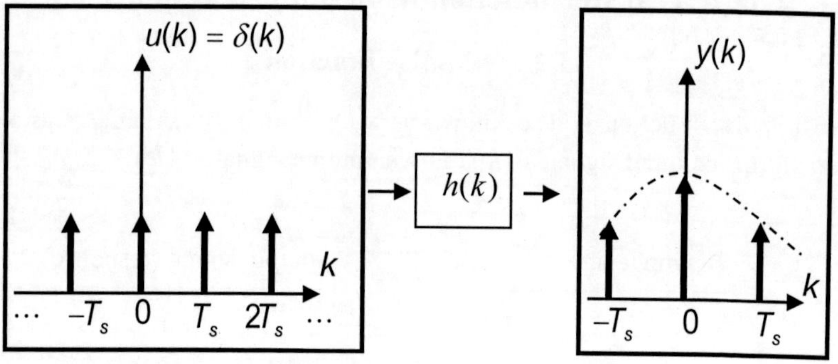

Fig. 3.4. The impulse response of a system

Note: $h(k)=0$ means the impulse response; if the system is causal, then $h(k)=0$ for $k<0$.

The response of a given linear time invariant system to an input $u(k)$ is the output $y(k)$.

$$y[k]=h[k]\otimes u[k]=h[k]\otimes \delta[k]$$

We recall the third property of z-transforms above that converts convolutions to multiplications as given by the following expression.

$$\begin{aligned} Z\{y[k]\} &= Z\{h[k] \otimes u[k]\} \\ &= Z\{h[k] \otimes \delta[k]\} \\ &= Z\{h[k]\}.Z\{\delta[k]\} \\ &= H(z) \end{aligned}$$

The transfer function of a linear time invariant system is simply the z-transform of its impulse response:

$$H(z) = Z\{h[k]\} = h[0] + h[0]z^{-1} + h[2]z^{-2} + \ldots$$

Example: Given that the response of a linear time invariant system is as follows:

$$y(k) = 10y(k-1) + 5y(k-2) + 3u(k) + 2u(k-1)$$

by rearranging the expression

$$y(k) - 10y(k-1) - 5y(k-2) = 3u(k) + 2u(k-1)$$

Taking the z-transforms on both sides, we get

$$Z\{y(k) - 10y(k-1) - 5y(k-2)\} = Z\{3u(k) + 2u(k-1)\}$$

So we get

$$Y(z) - 10z^{-1}Y(z) - 5z^{-2}Y(z) = 3U(z) + 2z^{-1}U(z)$$

and the transfer function $H(z)$ is

$$\begin{aligned} H(z) &= \frac{Y(z)}{U(z)} \\ &= \frac{3 + 2z^{-1}}{1 - 10z^{-1} - 5z^{-2}} \end{aligned}$$

Finally, the transfer function is

$$H(z) = \frac{3z^2 + 2z}{z^2 - 10z - 5}$$

3.2.3. Calculating frequency response from transfer function

As for continuous-time systems, the frequency response of a discrete-time system can be calculated from the transfer function $H(z)$.

Assuming that the input signal exciting the system is the sinusoid

$$u[t_k] = U\sin(wt_k) = U\sin(wkT_s)$$

where w is the signal frequency in rad/s, it can be shown that the stationary response on the output of the system is

$$\begin{aligned} y(t_k) &= Y\sin(wkT_s + \phi) \\ &= UA\sin(wkT_s + \phi) \\ &= \underbrace{U\overbrace{\left|H(e^{jwT_s})\right|}^{A}}_{Y}\sin\left[wt_k + \underbrace{\arg H(e^{jwT_s})}_{\phi}\right] \end{aligned}$$

where $H(e^{jwT_s})$ is the frequency response which is calculated with the following substitution:

$$H(e^{jwT_s}) = H(z)\big|_{z=e^{jwT_s}}$$

where T_s is the time-step.

The amplitude gain function is

$$A(w) = \left|H(e^{jwT_s})\right|$$

The phase lag function is

$$\phi(w) = \arg H(e^{jwT_s})$$

$A(w)$ and $\phi(w)$ can be plotted in a Bode diagram.

Example 1

Let us consider a continuous-time linear time invariant system given by the following transfer function:

$$H(s) = \frac{-1}{1-s}$$

Using the MatLab function and the sampling period $T_s = 0.2s$, we write

```
>> num=[-1];
>> den=[-1 1];
>>Hs=tf(num,den)
Hs =
    1
  -----
  s - 1
 Continuous-time transfer function.
>> Hz=c2d(Hs,0.2)
Hz =

   0.2214
  ---------
  z - 1.221

 Sample time: 0.2 seconds
 Discrete-time transfer function.
```

The discrete-time transfer function is

$$H(z)=\frac{b}{z-a}=\frac{0.2214}{z-1.221}$$

Using the following MatLab function

```
>> bode(Hz),grid
```

Fig. 3.5 shows the example of the Bode plot of the frequency response of the transfer function.

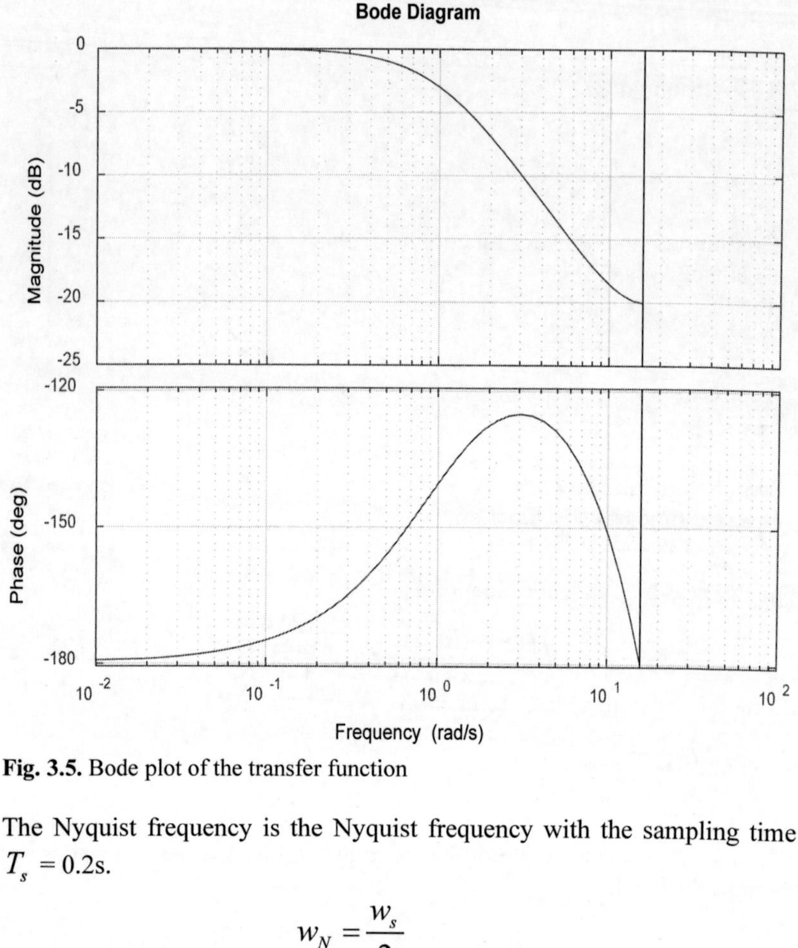

Fig. 3.5. Bode plot of the transfer function

The Nyquist frequency is the Nyquist frequency with the sampling time $T_s = 0.2$s.

$$
\begin{aligned}
w_N &= \frac{w_s}{2} \\
&= \frac{\pi}{T_s} \\
&= \frac{\pi}{0.2} \\
&\approx 15.7 rad/s
\end{aligned}
$$

The plots are not drawn above the Nyquist frequency because of the symmetry of the frequency response. This will be explained in the upcoming section.

Example 2

Calculating the frequency response algebraically from the z-transfer function

$$H(z) = \frac{b}{z-a}$$

the frequency response becomes

$$H(z = e^{jwT_s}) = \frac{b}{e^{jwT_s} - a}$$

$$= \frac{b}{\cos(wT_s) + j\sin(wT_s) - a}$$

$$= \frac{b}{\cos(wT_s) - a + j\sin(wT_s)}$$

$$= \frac{b}{\sqrt{\left(\cos(wT_s) - a\right)^2 + \left(\sin(wT_s)\right)^2}\, e^{j\arctan\frac{\sin(wT_s)}{\cos(wT_s)-a}}}$$

$$H(z = e^{jwT_s}) = \frac{b}{\sqrt{\left(\cos(wT_s) - a\right)^2 + \left(\sin(wT_s)\right)^2}}\, e^{-j\arctan\frac{\sin(wT_s)}{\cos(wT_s)-a}}$$

The amplitude gain function is

$$A(w) = \left|H(e^{jwT_s})\right| = \frac{b}{\sqrt{\left(\cos(wT_s) - a\right)^2 + \left(\sin(wT_s)\right)^2}}$$

and the phase lag function is

$$\phi(w) = \arg H(e^{jwT_s}) = -\arctan\frac{\sin(wT_s)}{\cos(wT_s) - a} \qquad [rad]$$

3.3. Transfer function from the difference equation

3.3.1 Difference equation models

The differential equation is the basic model type of continuous in time dynamic systems. However, the difference equation is the basic model type of discrete in time dynamic systems.

Here is an example of a linear first order difference equation with u as an input variable and y as an output variable:

$$y(k) = -a_1 y(k-1) + b_0 u(k) + b_1 u(k-1)$$

where a_1, b_0 and b_1 are coefficients of the difference equation. This difference equation is normalized, since the coefficient of $y(k)$ is 1 and may be written in an equivalent form as

$$y(k+1) + a_1 y(k) = b_0 u(k+1) + b_1 u(k)$$

where there are no time delayed terms, only time advanced terms (or terms without any advance or delay).

3.3.2 The z-transfer function

The general form of a linear finite difference equation with a constant-coefficient that describes a discrete linear time invariant system is given by the following equation:

$$y(k) + a_1 y(k-1) + \cdots + a_N y(k-N) = b_0 u(k) + b_1 u(k-1) + \cdots + b_M u(k-M)$$

It may be written in the compact form as:

$$y(k) + \sum_{i=1}^{N} a_i y(k-i) = \sum_{j=0}^{M} b_j u(k-j)$$

with M and N being positive integers and $a_1, \ldots, a_N$; $b_0, \ldots., b_M$ constant coefficients.

The discrete-time transfer function can be found by applying linearity and shift properties, taking the z-transform of both sides of the above:

$$Z\left[y(k)\right] + \sum_{i=1}^{N} a_i Z\left[y(k-i)\right] = \sum_{j=0}^{M} b_j Z\left[u(k-j)\right]$$

which leads to

$$Y(z)+\sum_{i=1}^{N}a_i z^{-i}Y(z)=\sum_{j=0}^{M}b_j z^{-j}U(z)$$

So

$$Y(z)\left[1+\sum_{i=1}^{N}a_i z^{-i}\right]=U(z)\sum_{j=0}^{M}b_j z^{-j}$$

Finally, if $a_0=1$, the transfer function $H(z)$ is a rational function given by:

$$H(z)=\frac{Y(z)}{U(z)}=\frac{\sum_{k=0}^{M}b_k z^{-k}}{1+\sum_{k=1}^{N}a_k z^{-k}}=\frac{\sum_{k=0}^{M}b_k z^{-k}}{\sum_{k=0}^{N}a_k z^{-k}}$$

Example

Consider a discrete-time system described by the following difference equation:

$$y(k)+12y(k-1)+3y(k-2)=10u(k)+5u(k-1)+2u(k-2)$$

The transfer function $H(z)$ can be found by applying linearity and shift properties taking the z-transform on both sides of the above:

$$H(z)=\frac{Y(z)}{U(z)}=\frac{10+5z^{-1}+2z^{-2}}{1+12z^{-1}+3z^{-2}}=\frac{10z^2+5z+2}{z^2+12z+3}$$

Hence, z-transfer functions can be written with both positive and negative exponents of z.

3.3.3 Difference equation representation

In this section, we express the difference equation by using the q^{-1} operator as given in the following equation:

$$\begin{aligned}y(k)\left(1+a_1q^{-1}+\cdots+a_Nq^{-N}\right)&=q^{-d}\left(b_0+b_1q^{-1}+\cdots+b_Mq^{-M}\right)u(k)\\&=\left(b_0+b_1q^{-1}+\cdots+b_Mq^{-M}\right)u(k-d)\end{aligned}$$

Now, using the two polynomials

$$A(q^{-1}) = 1 + a_1 q^{-1} + \cdots + a_N q^{-N}$$

and

$$B(q^{-1}) = b_0 + b_1 q^{-1} + \cdots + b_M q^{-M}$$

the difference equation can be written as:

$$A(q^{-1}) y(k) = q^{-d} B(q^{-1}) u(k) = B(q^{-1}) u(k-d)$$

3.3.4 From the z-transfer functions to the difference equation

In the inverse operation, we can find the difference equation from a given z-transfer function. The difference equation is applied for instance in a filtering algorithm which is derived from a filtering transfer function or in a control function which is derived from a given controller transfer function.

Let us consider the transfer function $H(z)$ of a discrete-time system which is given by the following equation:

$$H(z) = \frac{Y(z)}{U(z)} = \frac{\sum_{k=0}^{M} b_k z^{-k}}{1 + \sum_{k=1}^{N} a_k z^{-k}}$$

from where

$$Y(z)\left[1 + \sum_{k=1}^{N} a_k z^{-k}\right] = U(z)\left[\sum_{k=0}^{M} b_k z^{-k}\right]$$

Then

$$Y(z) + a_1 z^{-1} Y(z) + \ldots + a_N z^{-N} Y(z) = b_0 U(z) + b_1 z^{-1} U(z) + \ldots + b_M z^{-M} U(z)$$

By applying the inverse z-transform, the difference equation is given by the following expression:

$$y(k) + a_1 y(k-1) + \cdots + a_N y(k-N) = b_0 u(k) + b_1 u(k-1) + \cdots + a_M u(k-M)$$

Example

Consider a discrete-time system described by the transfer function

$$H(z)=\frac{Y(z)}{U(z)}=\frac{1+2z^{-1}}{1+z^{-1}+z^{-2}}$$

then

$$Y(z)\left[1+z^{-1}+z^{-2}\right]=U(z)\left[1+2z^{-1}\right]$$

which leads to the difference equation by applying the inverse z-transform

$$y(k)+y(k-1)+y(k-2)=u(k)+2u(k-1)$$

3.4 Rational z-transforms

The discrete-time z-transfer function $H(z)$ is given by

$$H(z)=\frac{N(z)}{D(z)}=\frac{\prod_k (z-z_k)}{\prod_k (z-p_k)}$$

where p_k and z_k are the poles and zeros of the z-transfer function. The z-transform of the unit sample response $h(k)$, denoted by $H(z)=Z\left\{h(k)\right\}$, is rational.

3.4.1 Poles and zeros

The zeros of a z-transfer function $H(z)$ are the values of z where $N(z)=0$. However, the poles of a z-transfer function $H(z)$ are the values of z where $D(z)=0$ and they are defined as the system poles.

Note that

$$\lim_{z\to z_k} H(z)=0 \text{ and } \lim_{z\to p_k} H(z)=\infty$$

If $H(z)$ is a rational function, *i.e.*, a fraction of two polynomials in z, then

$$H(z)=\frac{N(z)}{D(z)}=\frac{b_0+b_1z^{-1}+...+b_Mz^{-M}}{a_0+a_1z^{-1}+...+a_Nz^{-N}}=\frac{\sum_{k=0}^{M}b_kz^{-k}}{\sum_{k=0}^{N}a_kz^{-k}}$$

We assume $a_0 \neq 0$ and $b_0 \neq 0$, so we can rewrite

$$H(z)=\frac{b_0}{a_0}\frac{z^{-M}}{z^{-N}}\frac{z^M+\frac{b_1}{b_0}z^{M-1}+...+\frac{b_M}{b_0}}{z^N+\frac{a_1}{a_0}z^{N-1}+...+\frac{a_N}{a_0}}$$

$$=\frac{b_0}{a_0}z^{N-M}\frac{(z-z_1)(z-z_2)...(z-z_M)}{(z-p_1)(z-p_2)...(z-p_N)}$$

$$=H_0z^{N-M}\frac{\prod_{k=1}^{M}(z-z_k)}{\prod_{k=1}^{N}(z-p_k)}$$

With $H_0=\frac{b_0}{a_0}$, $H(z)$ has M finite zeros at $z_1,...,z_M$ and N finite poles at $p_1,...,p_N$.

$H_0=\frac{b_0}{a_0}$ is the scale factor which affects the amplitude of the system, whereas the poles and zeros affect its behavior.

Example

Let us consider the following system given by its transfer function

$$H(z)=\frac{N(z)}{D(z)}=\frac{3}{1+3z^{-1}+2z^{-2}}$$

$$H(z) = 3\frac{1}{z^{-2}}\frac{1}{z^2+2z+0.5}$$
$$= 3z^2 \frac{1}{(z+2)(z+1)}$$

We can use MatLab's function tf2zp which finds the zeros, poles, and gains of a rational function.

```
>> num=[3];
>> den=[1 3 2];
>> [zeros,poles,gains]=tf2zp(num,den)
zeros =
  Empty matrix: 0-by-1
poles =
   -2
   -1
gains =
    1
```

3.4.2 Pole and zero location and time-domain behavior

Once the poles and zeros have been found for a given z-transfer function $H(z)$, they can be plotted onto the z-plane. The z-plane is a complex plane with an imaginary and real axis referring to the complex variable z.

The position on the complex plane is given by $re^{i\theta}$ and the angle from the positive, real axis around the plane is denoted by θ. When mapping poles and zeros onto the plane, poles are denoted by an "$\times$" and zeros by an "o".

The below figure shows the z-plane. Some examples of plotting zeros and poles onto the plane can be found in the following section.

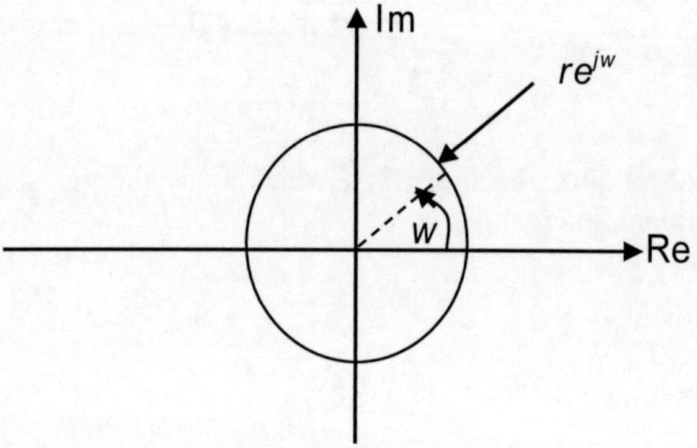

Fig. 3.6. The z-plane of the z-transfer function

Example

This section lists an example of finding the poles and zeros of a transfer function and then plotting them onto the z-plane.

$$H(z) = \frac{(z-0.2)(z+0.7)}{(z-0.29)(z+0.5)}$$

- The **zeros** of a z-transfer function $H(z)$ are: $\{0.2;-0.7\}$.
- The **poles** of a z-transfer function $H(z)$ are: $\{-0.29;-0.5\}$.

Computing the zeros and poles found from the transfer function by the Matlab z-plane function, the two zeros, mapped at -0.7 and 0.2, and the two poles, placed at 0.29 and -0.5, are plotted in the following figure.

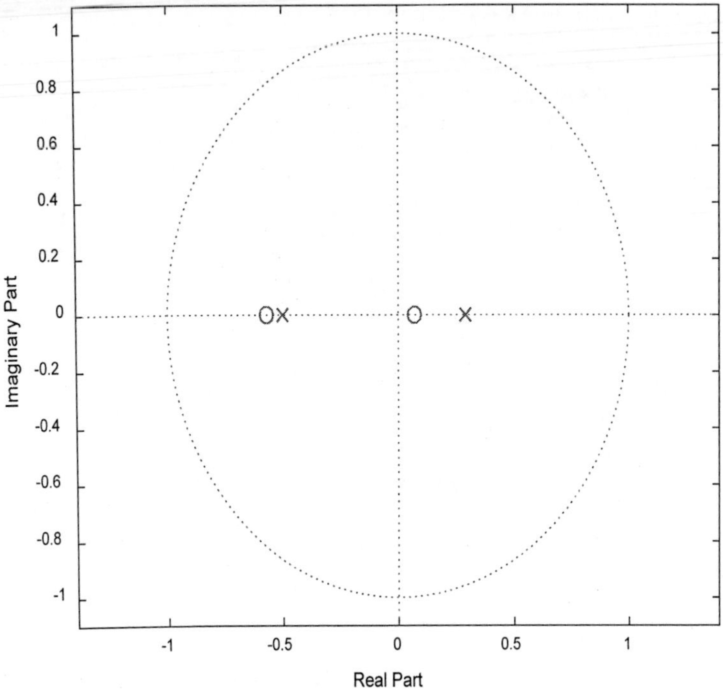

Fig. 3.7. The z-plane of the z-transfer function

3.5. Transfer function of a continuous system follows a zero-order hold

If the continuous in time transfer function H(s) follows a zero-order hold, see Fig. 3.8.

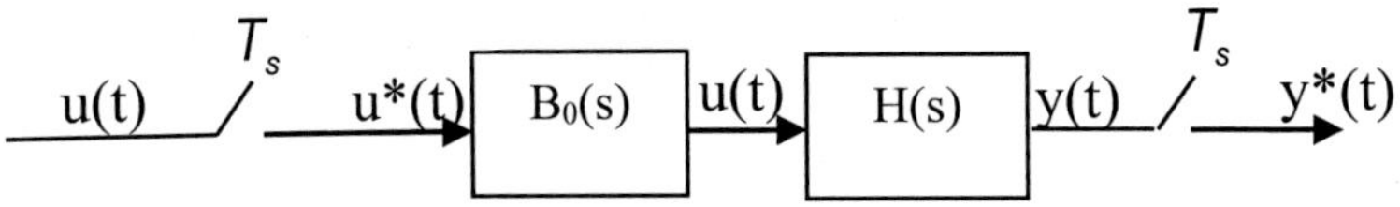

Fig. 3.8. The system follows a zero-order hold

where the transfer function of the zero-order hold is

$$B_0(s) = \frac{1 - e^{-sT_s}}{s}$$

The continuous in time system output is

$$Y(s)=H(s)B_0(s)U^*(s)$$

and the sampled system output is

$$Y^*(s)=\left(H(s)B_0(s)U^*(s)\right)^*=\left(H(s)B_0(s)\right)^*U^*(s)$$

However

$$Y(z)=Y^*(s)$$
$$U(z)=U^*(s)$$

So

$$Y(z)=Z\left[H(s)B_0(s)\right]U(z)$$

and the transfer function is

$$H(z)=\frac{Y(z)}{U(z)}=Z\left[H(s)B_0(s)\right]$$

Replacing the zero-order hold by its transfer function, the whole expression becomes

$$\begin{aligned}H(z)&=Z\left[H(s)\frac{1-e^{-sT_s}}{s}\right]\\&=Z\left[\frac{H(s)}{s}-e^{-sT_s}\frac{H(s)}{s}\right]\\&=Z\left[\frac{H(s)}{s}\right]-Z\left[e^{-sT_s}\frac{H(s)}{s}\right]\\&=Z\left[\frac{H(s)}{s}\right]-z^{-1}Z\left[\frac{H(s)}{s}\right]\\&=(1-z^{-1})Z\left[\frac{H(s)}{s}\right]\end{aligned}$$

In the case of a shifted system, the transfer function is

$$H(z)=(1-z^{-1})Z\left[\frac{H(s)}{s}e^{-\tau s}\right]$$

where $\tau=kT_s$ and then the discrete transfer function of the shifted system is

$$H(z) = z^{-k}(1 - z^{-1})Z\left[\frac{H(s)}{s}\right]$$

3.6. Transfer function of block diagrams

It is often convenient to decompose a linear system into small sub-systems that are interconnected. Moreover, all discrete systems can be built from addition, multiplication, duplication and delay modules. Hence, it is practical to know how to compute a transfer function from a block diagram and vice versa.

Most of these formulae follow directly from the definitions, whereas the loop-back construction is a bit trickier to analyse. As usual, let u and y be the input and the output signals. Additionally, let v be the feedback signal. Then their corresponding z-transforms must satisfy the following constraints.

3.6.1. Rules for block diagram manipulation

The most common rules for block diagram manipulation are shown. The rules are the same as for Laplace transfer functions. In the following figure the splitting sum-junction is presented.

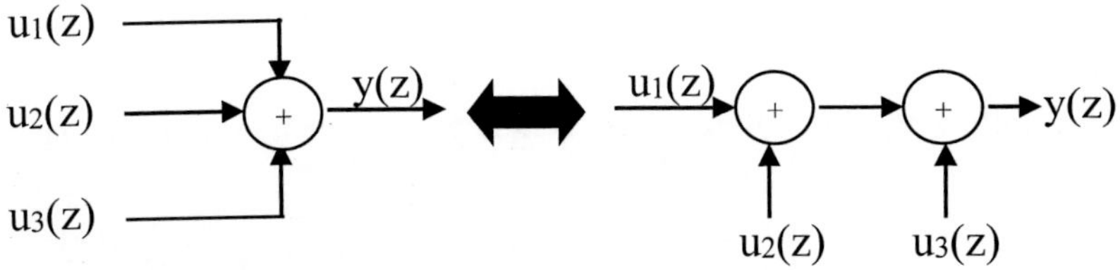

Fig. 3.9. Splitting sum junction

In the following figure the moving sum-junction is presented.

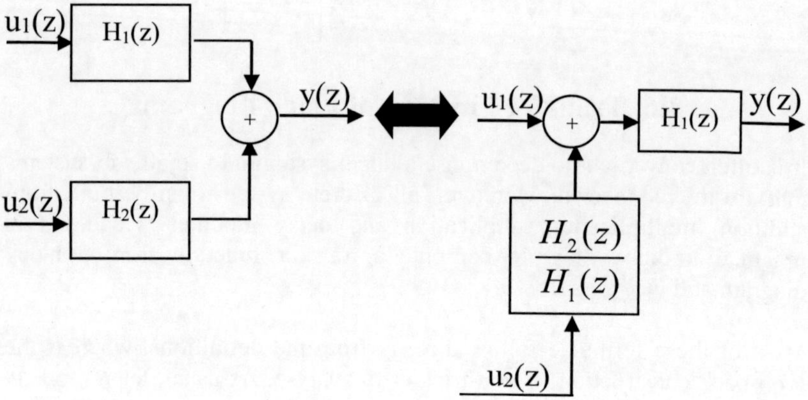

Fig. 3.10. Moving sum junction

In the following figure the moving branch is presented.

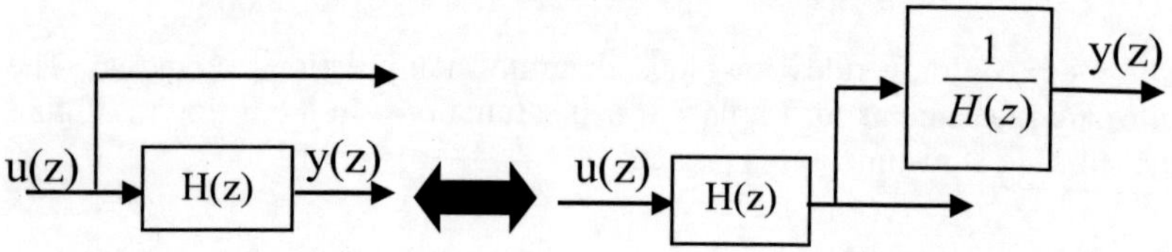

Fig. 3.11. Moving branch

The following figure represents the moving branch across the sum-junction.

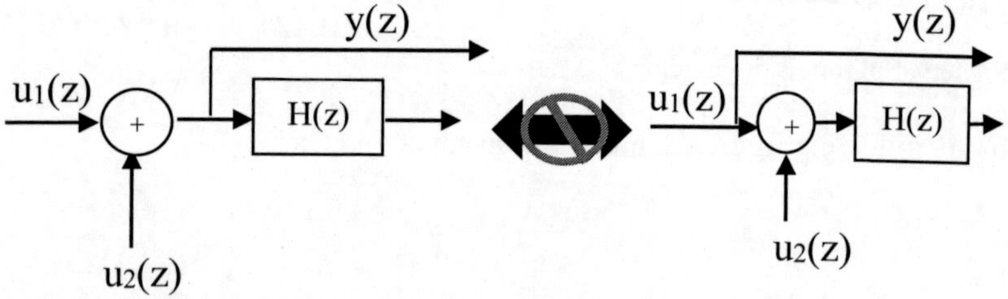

Fig. 3.12. Moving branch across the sum junction

3.6.2. Influence of the position's sampler

The absence or presence of an input sampler is crucial in determining the transfer function of a system. Let us consider that a sampler is put in the input of the continuous system $H(s)$ where $U(s)$ and $Y(s)$ are respectively the Laplace transforms of $u(t)$ and $y(t)$ - see Fig. 3.13.

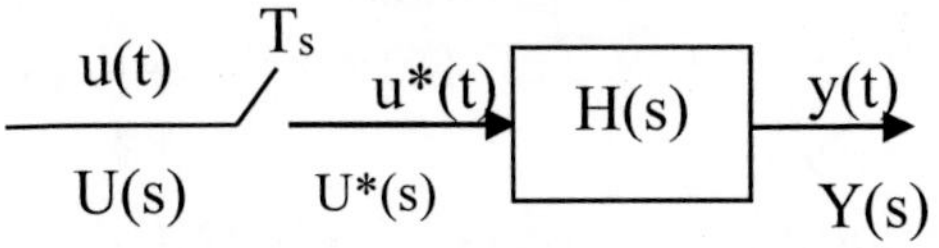

Fig. 3.13. The sampler is located before the system

$Y(s)$ is computed as

$$Y(s) = H(s)U^*(s)$$

$Y^*(s)$ is given as

$$Y^*(s) = \left(H(s)U^*(s)\right)^* = H^*(s)U^*(s)$$

or

$$Y(z) = Y^*(s)$$
$$U(z) = U^*(s)$$

so

$$Y(z) = H(z)U(z)$$

In the second figure, the sampler is placed in the output of the system - see Fig. 3.14.

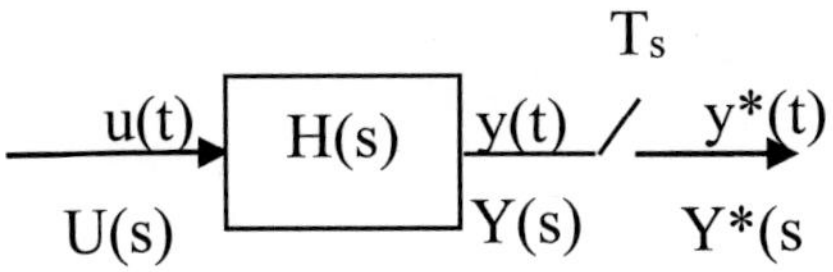

Fig. 3.14. The sampler is located after the system

$Y(s)$ is computed as

$$Y(s) = H(s)U(s)$$

$Y^*(s)$ is given as

$$Y^*(s) = \left(H(s)U(s)\right)^*$$

So

$$Y(z) = \left(H(s)U(s)\right)^* = Z\left[H(s)U(s)\right]$$

From Figures 3.13 and 3.14, we can deduce that

$$Y(z) = H(z)U(z)Y(z) \neq Z\left[H(s)U(s)\right]$$

3.6.3. Multiplied coefficient of discrete-time systems

Let α be a multiplied coefficient of discrete-time systems as shown in the following figure.

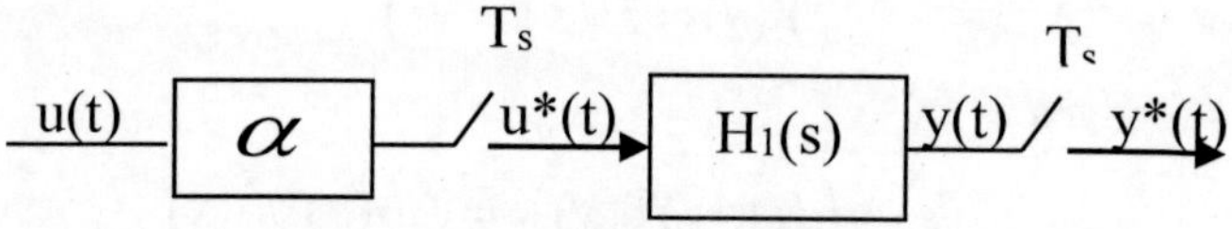

Fig. 3.15. A coefficient is multiplied to the sampled system

The transfer function is

$$H(z) = \frac{Y(z)}{U(z)} = \alpha H_1(z)$$

3.6.4. Addition of discrete-time systems

Let us consider two discrete-time systems in parallel connection as shown by the following figure:

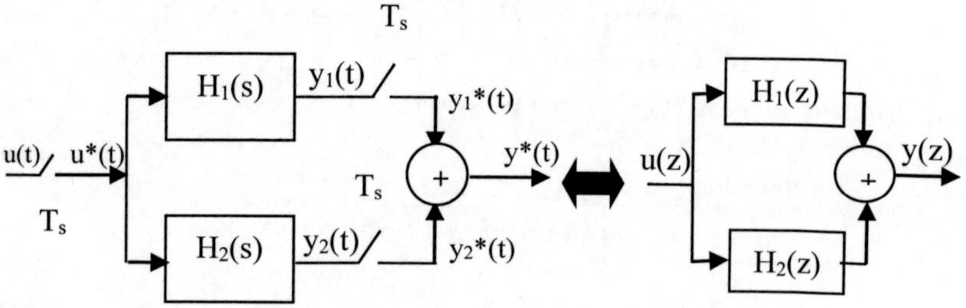

Fig. 3.16. Addition of two discrete-time systems

The first output $Y_1(s)$ is

$$Y_1(s) = H_1(s)U^*(s)$$

The sampled first output $Y_1^*(s)$ is

$$Y_1^*(s) = \left(H_1(s)U^*(s)\right)^* = H_1^*(s)U^*(s)$$

The z-transform $Y_1(z)$ is

$$Y_1(z) = H_1(z)U(z)$$

The second output $Y_2(s)$ is

$$Y_2(s) = H_2(s)U^*(s)$$

The sampled second output $Y_2^*(s)$ is

$$Y_2^*(s) = \left(H_2(s)U^*(s)\right)^* = H_2^*(s)U^*(s)$$

The z-transform $Y_2(z)$ is

$$Y_2(z) = H_2(z)U(z)$$

Then

$$Y(z) = Y_1(z) + Y_2(z) = \left(H_1(z) + H_2(z)\right)U(z)$$

Finally, the z-transfer function is

$$H(z) = \frac{Y(z)}{U(z)} = H_1(z) + H_2(z)$$

In general

Let us consider n discrete-time systems in parallel connection as shown by the following figures.

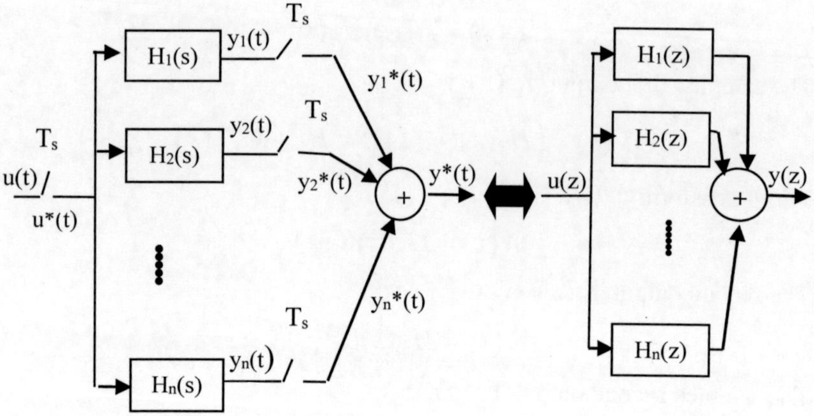

Fig. 3.17. Addition of n discrete-time systems

The transfer function is

$$H(z) = \frac{Y(z)}{U(z)} = H_1(z) + H_2(z) + \ldots + H_n(z) = \sum_{i=1}^{n} H_i(z)$$

3.6.5. Cascaded separated continuous-time systems

Let us consider two continuous systems in series connection separated by a sampler of sampling period T_s.

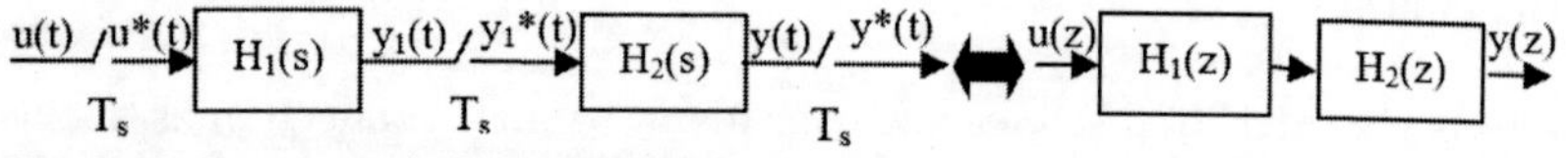

Fig. 3.18. Two cascaded separated continuous-time systems

The output of the second block $Y(s)$ is

$$Y(s) = H_2(s) Y_1^*(s)$$

The sampled output of the second block $Y^*(s)$ is

$$Y^*(s) = \left(H_2(s) Y_1^*(s)\right)^* = H_2^*(s) Y_1^*(s)$$

The z-transform $Y(z)$ is

$$Y(z) = H_2(z)Y_1(z) \qquad (*)$$

or

$$Y_1(s) = H_1(s)U^*(s)$$

or the sampled expression is

$$Y_1^*(s) = \left(H_1(s)U^*(s)\right)^* = H_1^*(s)U^*(s)$$

so

$$Y_1(z) = H_1(z)U(z) \qquad (**)$$

Using (**) in (*), thus

$$Y(z) = H_2(z)H_1(z)U(z)$$

Finally, the transfer function is given as

$$H(z) = \frac{Y(z)}{U(z)} = H_1(z)H_2(z)$$

In general:

Let n continuous systems in series be separated by a sampler of sampling period T_s.

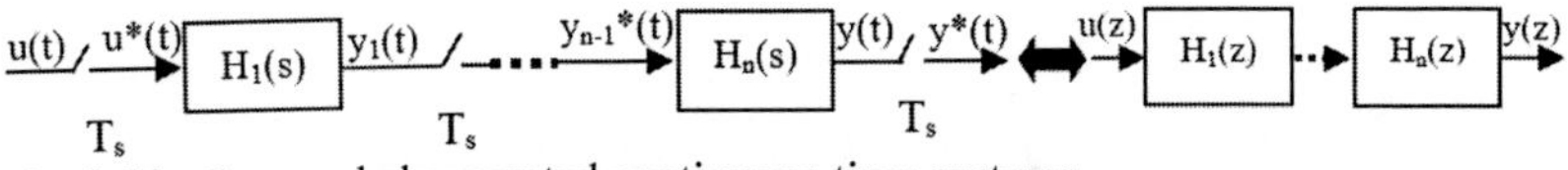

Fig. 3.19. n cascaded separated continuous-time systems

The transfer function is

$$H(z) = \frac{Y(z)}{U(z)} = H_1(z)H_2(z)\dots H_n(z) = \prod_{i=1}^{n} H_i(z)$$

3.6.6. Cascaded non separated continuous-time systems

Let us consider two continuous systems in series that are not separated by a sampler of sampling period T_s.

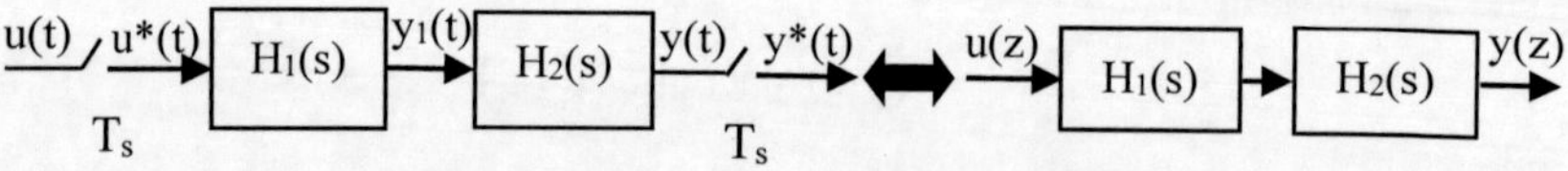

Fig. 3.20. Two cascaded continuous-time systems not separated

The output of the second block $Y(s)$ is

$$Y(s) = H_2(s)Y_1(s) \Rightarrow Y(s) = H_2(s)H_1(s)U^*(s)$$

The sampled output of the second block $Y^*(s)$ is

$$Y^*(s) = \left(H_2(s)H_1(s)U^*(s)\right)^* = \left(H_2(s)H_1(s)\right)^* U^*(s)$$

and

$$Y(z) = Z\left[H_2(s)H_1(s)\right]U(z)$$

Finally, the transfer function is

$$H(z) = \frac{Y(z)}{U(z)} = Z\left[H_2(s)H_1(s)\right]$$

In general:

Let us consider n continuous systems in series that are not separated by a sampler of sampling period T_s.

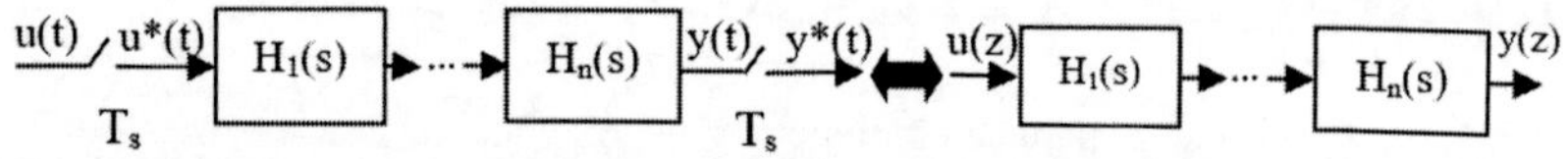

Fig. 3.21. n cascaded non separated continuous-time systems

The transfer function is

$$H(z) = \frac{Y(z)}{U(z)} = Z\left[H_1(s)H_2(s)\dots H_n(s)\right] = Z\left[\prod_{i=1}^{n} H_i(s)\right]$$

3.7. Transfer function of closed-loop discrete-time systems

Discrete-time systems are conveniently described by block diagrams and transfer functions can be determined from them.

3.7.1. Case of two systems that are not separated by a sampler

In this case of a closed-loop system the sampler is only used before the system defined by $H_1(s)$ - see Fig. 3.22.

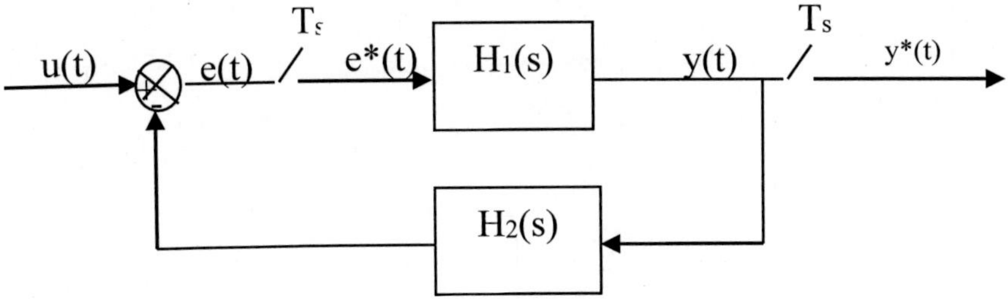

Fig. 3.22. Two systems that are not separated by a sampler

The system output is

$$Y(s)=H_1(s)E^*(s)$$

The sampled system output is

$$Y^*(s)=\left(H_1(s)E^*(s)\right)^*=H_1^*(s)E^*(s)$$

The z-transform of the system output is

$$Y(z)=H_1(z)E(z) \qquad (*)$$

The second variable is given by the following expression

$$E(s)=U(s)-H_2(s)H_1(s)E^*(s)$$

The sampled variable is

$$E^*(s)=U^*(s)-\left(H_2(s)H_1(s)E^*(s)\right)^*=U^*(s)-\left(H_2(s)H_1(s)\right)^*E^*(s)$$

The z-transform of the variable is

$$E(z)=U(z)-Z\left[H_2(s)H_1(s)\right]E(z)=\frac{U(z)}{1+Z\left[H_2(s)H_1(s)\right]} \qquad (**)$$

Using (**) into (*), the z-transform of the output is

$$Y(z)=\frac{H_1(z)U(z)}{1+Z\left[H_2(s)H_1(s)\right]}$$

Finally, the z-transfer function is

$$H(z) = \frac{Y(z)}{U(z)} = \frac{H_1(z)}{1 + Z\left[H_2(s)H_1(s)\right]}$$

3.7.2. Case of two systems that are separated by a sampler

In this case of a closed-loop system, two samplers are used and the system is defined by $H_1(s)$ - see Fig. 3.23.

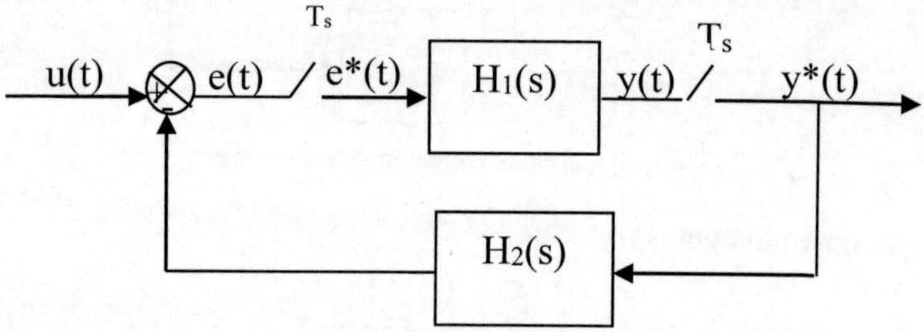

Fig. 3.23. Two systems that are separated by a sampler

The system output is

$$Y(s) = H_1(s)E^*(s) \qquad (*)$$

The sampled system output is

$$Y^*(s) = \left(H_1(s)E^*(s)\right)^* = H_1^*(s)E^*(s) \qquad (**)$$

The z-transform of the system output is

$$Y(z) = H_1(z)E(z)$$

The second variable is given by the following expression

$$E(s) = U(s) - H_2(s)H_1^*(s)E^*(s)$$

The sampled variable is

$$E^*(s) = U^*(s) - \left(H_2(s)H_1^*(s)E^*(s)\right)^* = U^*(s) - H_2^*(s)H_1^*(s)E^*(s)E^*(s)$$

The z-transform of the variable is

$$E(z) = U(z) - H_2(z)H_1(z)E(z) = \frac{U(z)}{1 + H_2(z)H_1(z)} \qquad (***)$$

Using (***) into (**) and (*), the z-transform of the system output is

$$Y(z)=\frac{H_1(z)U(z)}{1+H_2(z)H_1(z)}$$

Finally, the z-transfer function is

$$H(z)=\frac{Y(z)}{U(z)}=\frac{H_1(z)}{1+H_2(z)H_1(z)}$$

3.7.3. Case of two samplers and three systems

In this case of a closed-loop system, two samplers are applied and three systems are used - see Fig. 3.24.

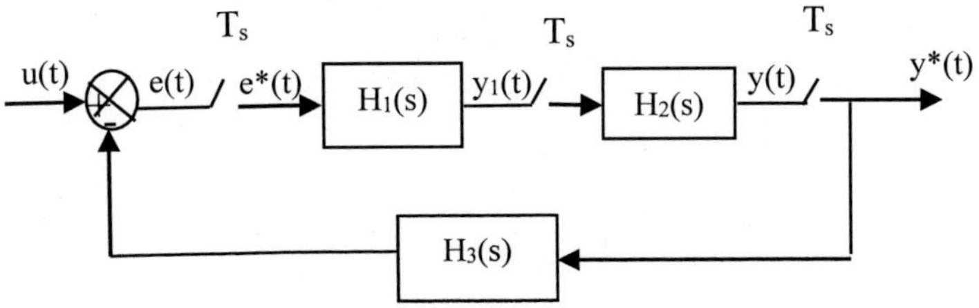

Fig. 3.24. Three systems and two samplers

The system output is

$$Y(s)=H_2(s)Y_1^*(s)$$

The sampled system output is

$$Y^*(s)=\left(H_2(s)Y_1^*(s)\right)^*=H_2^*(s)Y_1^*(s)$$

The z-transform of the system output is

$$Y(z)=H_2(z)Y_1(z) \qquad (*)$$

The first system output is

$$Y_1(s)=H_1(s)E^*(s)$$

The sampled first system output is

$$Y_1^*(s)=\left(H_1(s)E(s)\right)^*=H_1^*(s)E^*(s)$$

The z-transform of the system output is

$$Y_1(z)=H_1(z)E(z) \qquad (**)$$

Using the equation (**) into (*) we find

$$Y(z) = H_2(z)H_1(z)E(z)$$

Where $E(s)$ is given by the following expression

$$E(s) = U(s) - H_3(s)Y(s) = U(s) - H_3(s)H_2(s)H_1^*(s)E^*(s)$$

the sampled $E(s)$ is

$$E^*(s) = U^*(s) - \left(H_3(s)H_2(s)H_1^*(s)E^*(s)\right)^* = U^*(s) - \left(H_3(s)H_2(s)\right)^* H_1^*(s)E^*(s)$$

and the z-transform function is

$$E(z) = U(z) - \left(H_3(s)H_2(s)\right)^* H_1(z)E(z) \Rightarrow E(z) = \frac{U(z)}{1 + Z\left(H_3(s)H_2(s)\right)H_1(z)} \quad (***)$$

Using (*) into (***), the z-transform is

$$Y(z) = \frac{H_1(z)H_2(z)U(z)}{1 + Z\left(H_3(s)H_2(s)\right)H_1(z)}$$

Finally, the z-transfer function is

$$H(z) = \frac{Y(z)}{U(z)} = \frac{H_1(z)H_2(z)}{1 + Z\left(H_3(s)H_2(s)\right)H_1(z)}$$

3.7.4. Case of a sampler and three systems

In this case of a closed-loop system a sampler is applied and three systems are used - see Fig. 3.25.

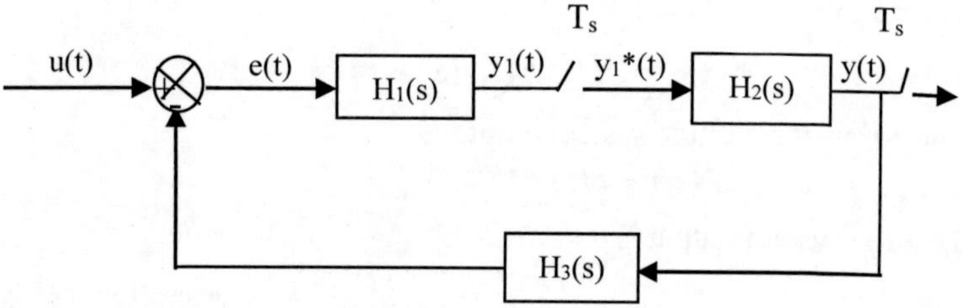

Fig. 3.25. Three systems and a sampler

The system output is

$$Y(s) = H_2(s)Y_1^*(s)$$

The sampled system output is

$$Y^*(s) = \left(H_2(s)Y_1^*(s)\right)^* = H_2^*(s)Y_1^*(s)$$

The z-transform of the system output is

$$Y(z) = H_2(z)Y_1(z) \qquad (*)$$

However,

$$E(s) = U(s) - H_3(s)Y(s)$$

and

$$\begin{aligned} Y_1(s) &= H_1(s)E(s) \\ &= H_1(s)U(s) - H_1(s)H_3(s)Y(s) \\ &= H_1(s)U(s) - H_1(s)H_3(s)H_2(s)H_1^*(s)Y_1^*(s) \end{aligned}$$

The sampled system output $Y_1^*(s)$ is

$$Y_1^*(s) = \left(H_1(s)E(s)\right)^* - \left(H_1(s)H_2(s)H_3(s)Y_1^*(s)\right)^* \qquad (**)$$

The z-transform of the system output is

$$Y_1(z) = Z\left(H_1(s)E(s)\right) - Z\left(H_1(s)H_2(s)H_3(s)\right)Y_1(z)$$

$$\Rightarrow Y_1(z) = \frac{Z\left(H_1(s)E(s)\right)}{1 + Z\left(H_1(s)H_2(s)H_3(s)\right)}$$

Finally, the z-transfer function is

$$Y(z) = H_2(z)Y_1(z) = \frac{H_2(z)Z\left(H_1(s)E(s)\right)}{1 + Z\left(H_1(s)H_2(s)H_3(s)\right)}$$

3.7.5. Case of a sampler and two systems

In this case of a closed-loop system, a sampler is applied and two systems are used - see Fig. 3.26.

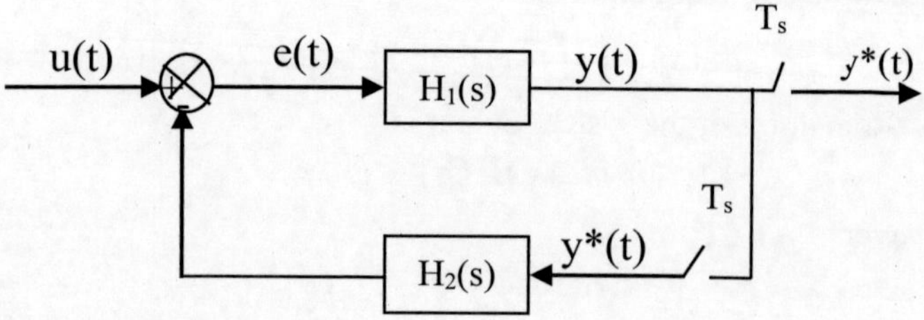

Fig. 3.26. Two systems and a sampler

We compute $E(s)$ as

$$E(s) = U(s) - H_2(s)Y^*(s)$$

yet we know that $Y(s)$ is given as

$$Y(s) = H_1(s)E(s) = H_1(s)U(s) - H_1(s)H_2(s)Y^*(s)$$

and the sampled system output is given as

$$Y^*(s) = \left(H_1(s)E(s)\right)^* - \left(H_1(s)H_2(s)\right)^* Y^*(s)$$

Then, the z-transform is

$$Y(z) = Z\left(H_1(s)E(s)\right) - Z\left(H_1(s)H_2(s)\right)Y(z)$$

Thus

$$Y_1(z) = \frac{Z\left(H_1(s)E(s)\right)}{1 + Z\left(H_1(s)H_2(s)\right)}$$

3.8. State space description and z-transfer function

3.8.1. Definition

The state variable technique is used to convert transfer function system representations into some first order difference equations, which many advanced matrix theories and computational tools can be applied to. Thus, the control system design using the state variable technique can be done in a very systematic fashion and many well-developed commercial software tools, such as MatLab, can be readily and easily utilized. Solutions to the control problem can be found in a straightforward manner.

There are two important equations associated with such a technique, the state equation and the output equation, which completely characterize the properties of a linear system. Any linear time-invariant system can be represented by a state space model with four constant matrices regardless of its dynamical order. For such reasons, most advanced control theories are always developed in the state space setting.

3.8.2. State space representation of a 1st order discrete system

Consider the transfer function, in the z-domain, of a 1st order discrete system:

$$H(z)=\frac{Y(z)}{U(z)}=\frac{b_0}{z+a_0}$$

$$\Rightarrow (z+a_0)Y(z)=b_0U(z)$$

$$\Rightarrow y(k+1)+a_0y(k)=b_0u(k)$$

Define a so-called state variable as follows:

$$x(k)=y(k)$$

$$\Rightarrow x(k+1)=y(k+1)=-a_0y(k)+b_0u(k)=-a_0x(k)+b_0u(k)$$

$$\Rightarrow \begin{cases} x(k+1)=-a_0x(k)+b_0u(k) \\ y(k)=x(k) \end{cases}$$

$$\Rightarrow \begin{cases} x(k+1)=Ax(k)+Bu(k) \\ y(k)=Cx(k) \end{cases}$$

3.8.3. State space representation of a 2nd order discrete system

Consider the transfer function, in the z-domain, of a 2nd order discrete system:

$$H(z)=\frac{Y(z)}{U(z)}=\frac{b_0}{z^2+a_1z+a_0}$$

$$\Rightarrow (z^2+a_1z+a_0)Y(z)=b_0U(z)$$

$$\Rightarrow y(k+1)+a_1y(k+1)+a_0y(k)=b_0u(k)$$

Define a so-called state variable as follows:

$$\Rightarrow \begin{cases} x_1(k)=y(k) \\ x_2(k)=y(k+1) \end{cases}$$

$$\Rightarrow\begin{cases} x_1(k+1) = y(k+1) = x_2(k) \\ x_2(k+1) = y(k+2) = -a_1 y(k+1) - a_0 y(k) + b_0 u(k) \end{cases}$$

$$\Rightarrow\begin{cases} x_1(k+1) = y(k+1) = x_2(k) \\ x_2(k+1) = -a_1 x_2(k) - a_0 x_1(k) + b_0 u(k) \end{cases}$$

$$\Rightarrow\begin{cases} \begin{bmatrix} x_1(k+1) \\ x_2(k+1) \end{bmatrix} = \begin{bmatrix} 0 & 1 \\ -a_0 & -a_1 \end{bmatrix} \begin{bmatrix} x_1(k) \\ x_2(k) \end{bmatrix} + \begin{bmatrix} 0 \\ b_0 \end{bmatrix} u(k) \\ y(k) = x_1(k) = \begin{bmatrix} 1 & 0 \end{bmatrix} \begin{bmatrix} x_1(k) \\ x_2(k) \end{bmatrix} \end{cases}$$

$$\Rightarrow\begin{cases} x(k+1) = Ax(k) + Bu(k) \\ y(k) = Cx(k) \end{cases}$$

Another form of transfer function, in the z-domain, of a 2nd order discrete system is:

$$H(z) = \frac{Y(z)}{U(z)} = \frac{b_1 z + b_0}{z^2 + a_1 z + a_0}$$

$$\Rightarrow \left(z^2 + a_1 z + a_0\right) Y(z) = \left(b_1 z + b_0\right) U(z)$$

$$\Rightarrow y(k+1) + a_1 y(k+1) + a_0 y(k) = b_1 u(k+1) + b_0 u(k)$$

Define a so-called state variable

$$\Rightarrow\begin{cases} x_1(k) = y(k) \\ x_2(k) = y(k+1) - b_1 u(k) \end{cases}$$

$$\Rightarrow\begin{cases} x_1(k+1) = y(k+1) = x_2(k) + b_1 u(k) \\ x_2(k+1) = y(k+2) - b_1 u(k+1) = -a_1 y(k+1) - a_0 y(k) + b_1 u(k+1) + b_0 u(k) - b_1 u(k+1) \end{cases}$$

$$\Rightarrow\begin{cases} x_1(k+1) = y(k+1) = x_2(k) + b_1 u(k) \\ x_2(k+1) = -a_1 x_2(k) - b_1 u(k) - a_0 x_1(k) + b_0 u(k) \end{cases}$$

$$\Rightarrow\begin{cases} x_1(k+1) = y(k+1) = x_2(k) + b_1 u(k) \\ x_2(k+1) = -a_0 x_1(k) - a_1 x_2(k) + \left(b_0 - a_1 b_1\right) u(k) \end{cases}$$

$$\Rightarrow \begin{cases} \begin{bmatrix} x_1(k+1) \\ x_2(k+1) \end{bmatrix} = \begin{bmatrix} 0 & 1 \\ -a_0 & -a_1 \end{bmatrix} \begin{bmatrix} x_1(k) \\ x_2(k) \end{bmatrix} + \begin{bmatrix} b_1 \\ b_0 - a_1 b_1 \end{bmatrix} u(k) \\ y(k) = x_1(k) = \begin{bmatrix} 1 & 0 \end{bmatrix} \begin{bmatrix} x_1(k) \\ x_2(k) \end{bmatrix} \end{cases}$$

$$\Rightarrow \begin{cases} x(k+1) = Ax(k) + Bu(k) \\ y(k) = Cx(k) \end{cases}$$

3.8.4. State space representation of a nth order discrete system

Consider the transfer function, in the z-domain, of a nth order discrete system:

$$H(z) = \frac{Y(z)}{U(z)} = \frac{b_m z^m + b_{m-1} z^{m-1} + \cdots + b_0}{z^n + a_{n-1} z^{n-1} + \cdots + a_0}$$

Using the partial fraction technique, we can rewrite this general system as

$$H(z) = \frac{Y(z)}{U(z)} = \frac{b_{1,1} z + b_{0,1}}{z^2 + a_{1,1} z + a_{0,1}} + \cdots + \frac{b_{1,k} z + b_{0,k}}{z^2 + a_{1,k} z + a_{0,k}} + \frac{b_0}{z + a_0} + d$$

where

$$\frac{b_{1,1} z + b_{0,1}}{z^2 + a_{1,1} z + a_{0,1}} \Rightarrow \begin{cases} x_1(k+1) = A_1 x_1(k) + B_1 u(k) \\ y_1(k) = C_1 x_1(k) \end{cases}$$

$$\frac{b_{1,k} z + b_{0,k}}{z^2 + a_{1,k} z + a_{0,k}} \Rightarrow \begin{cases} x_k(k+1) = A_k x_k(k) + B_k u(k) \\ y_k(k) = C_k x_k(k) \end{cases}$$

$$\frac{b_0}{z + a_0} \Rightarrow \begin{cases} x_0(k+1) = A_0 x_0(k) + B_0 u(k) \\ y_0(k) = C_0 x_0(k) \end{cases}$$

The state space representation of the complicated system is then given by

$$\begin{cases} x(k+1) = Ax(k) + Bu(k) \\ y(k) = Cx(k) + Du(k) \end{cases}$$

with

$$x=\begin{bmatrix} x_0 \\ x_1 \\ \vdots \\ x_k \end{bmatrix}, A=\begin{bmatrix} A_0 & 0 & \cdots & 0 \\ 0 & A_1 & \cdots & 0 \\ \vdots & \vdots & \ddots & \vdots \\ 0 & 0 & \cdots & A_k \end{bmatrix}, B=\begin{bmatrix} B_0 \\ B_1 \\ \vdots \\ B_k \end{bmatrix}, D=d$$

and

$$C=\begin{bmatrix} C_0 & C_1 & \cdots & C_k \end{bmatrix}$$

Fortunately, the predefined function **tf2ss** of MatLab can do this for us.

Example

Convert the discrete transfer function in the following equation into a state space representation.

$$H(z)=\frac{0.5z+0.5}{(z-0.5)(z-0.9)}=\frac{0.5z+0.5}{z^2-1.4z+0.45}$$

Recall the formula we have derived:

$$\Rightarrow\begin{cases} \begin{bmatrix} x_1(k+1) \\ x_2(k+1) \end{bmatrix}=\begin{bmatrix} 0 & 1 \\ -a_0 & -a_1 \end{bmatrix}\begin{bmatrix} x_1(k) \\ x_2(k) \end{bmatrix}+\begin{bmatrix} b_1 \\ b_0-a_1b_1 \end{bmatrix}u(k) \\ y(k)=x_1(k)=\begin{bmatrix} 1 & 0 \end{bmatrix}\begin{bmatrix} x_1(k) \\ x_2(k) \end{bmatrix} \end{cases}$$

$$\Rightarrow\begin{cases} x(k+1)=Ax(k)+Bu(k) \\ y(k)=Cx(k) \end{cases}$$

with

$$A=\begin{bmatrix} 0 & 1 \\ -0.45 & 1.4 \end{bmatrix}, B=\begin{bmatrix} 0.5 \\ 1.2 \end{bmatrix}, C=\begin{bmatrix} 1 & 0 \end{bmatrix}, D=0$$

and

$$C=\begin{bmatrix} C_0 & C_1 & \cdots & C_k \end{bmatrix}$$

Given a state space description (previous equation) of a time-invariant system, it is possible to compute the transfer matrix $H(z)$ as follows

$$H(z) = C(zI - A)^{-1} B + D$$

We verify this by using the functions **tf2ss** and **ss2tf** in MatLab.

3.9. Examples of transfer function applications

Application 1

Find the transfer function in the case of an open loop and closed loop given by the following figure.

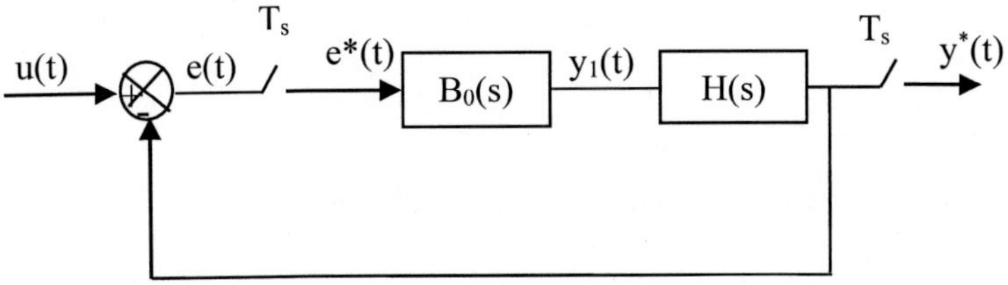

Fig. 3.27. A continuous system with zero-order-hold in the closed loop

Where the zero-order hold function is $B_0(s) = \dfrac{1 - e^{-sT_s}}{s}$ and the discrete-time system function is $H(s) = \dfrac{1}{s+1}$.

Answer

The transfer function of the open loop $H_{OL}(z)$ is

$$H_{OL}(z)=\left(1-z^{-1}\right)Z\left[\frac{H(s)}{s}\right]$$

$$=\left(1-z^{-1}\right)Z\left[\frac{1}{s(s+1)}\right]$$

$$=\left(1-z^{-1}\right)Z\left[\frac{1}{s}-\frac{1}{s+1}\right]$$

$$=\left(1-z^{-1}\right)\left(\frac{1}{1-z^{-1}}-\frac{1}{1-e^{-T_s}z^{-1}}\right)$$

$$=\frac{z^{-1}\left(1-e^{-T_s}\right)}{1-e^{-T_s}z^{-1}}$$

The transfer function of the closed loop $H_{CL}(z)$ is

$$H_{CL}(z)=\frac{Y(z)}{U(z)}$$

$$=\frac{H_{OL}(z)}{1+H_{OL}(z)}$$

$$=\frac{\dfrac{z^{-1}\left(1-e^{-T_s}\right)}{1-e^{-T_s}z^{-1}}}{1+\dfrac{z^{-1}\left(1-e^{-T_s}\right)}{1-e^{-T_s}z^{-1}}}$$

$$=\frac{z^{-1}\left(1-e^{-T_s}\right)}{1-e^{-T_s}z^{-1}+z^{-1}\left(1-e^{-T_s}\right)}$$

$$=\frac{z^{-1}\left(1-e^{-T_s}\right)}{1+\left(1-2e^{-T_s}\right)z^{-1}}$$

Application 2

Consider a continuous system with a transfer function $H(s)=\dfrac{K}{1+\tau s}$ followed by a zero-order-hold (ZOH) given by the following figure

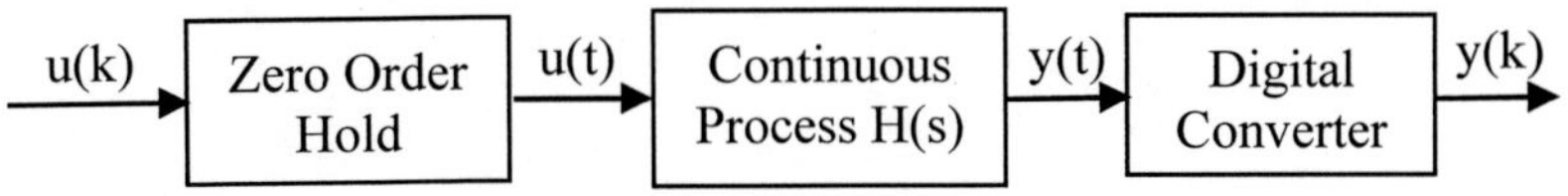

Fig. 3.28. A continuous system with a zero-order-hold in the open loop

1 - Find the z-transfer function $H(z)=\dfrac{\alpha}{z-z_1}$ of the continuous-time system $H(s)=\dfrac{\beta}{s-p_1}$ followed by the ZOH. Give α, β, z_1 and p_1 and compare the order of these transfer functions.

2 - Find the relation between the two poles of continuous-time p_1 and the discrete-time systems z_1.

3 - Check how one first order of a continuous system deals with many discrete-time systems.

Answer

1 - The transfer function of the continuous-time system is

$$H(s)=\frac{K}{1+\tau s}$$

$$H(s)=\frac{K\dfrac{1}{\tau}}{s+\dfrac{1}{\tau}}$$

$$=\frac{\beta}{s-p_1}$$

with $\beta=\dfrac{K}{\tau}$ and $p_1=-\dfrac{1}{\tau}$.

The z-transfer function is given by the following equation $H(z)$:

$$H(z) = (1 - z^{-1}) Z\left[\frac{H(s)}{s}\right]$$

$$H(z) = (1 - z^{-1}) Z\left[\frac{K}{s(1+\tau s)}\right]$$

$$= \frac{z-1}{z} \frac{Kz(1-e^{-\frac{T_s}{\tau}})}{(z-1)(z-e^{-\frac{T_s}{\tau}})}$$

$$H(z) = K\frac{(1-e^{-\frac{T_s}{\tau}})}{z-e^{-\frac{T_s}{\tau}}}$$

$$H(z) = \frac{\alpha}{z - z_1}$$

with $\alpha = K(1-e^{-\frac{T_s}{\tau}})$ and $z_1 = e^{-\frac{T_s}{\tau}}$.

Both systems have the same order $n = 1$.

2 - From the denominator of the continuous-time system, the pole is $p_1 = -\frac{1}{\tau}$ and from the denominator of the discrete-time system, the pole is $z_1 = e^{-\frac{T_s}{\tau}} = e^{p_1 T_s}$.

3 - If the sampling rate is $T_s = T_{s1}$, then

$$H(z) = \frac{K(1-e^{-\frac{T_{s1}}{\tau}})}{z-e^{-\frac{T_{s1}}{\tau}}}$$

If the sampling rate is $T_s = T_{s2}$, then

$$H(z) = \frac{K(1-e^{-\frac{T_{s2}}{\tau}})}{z-e^{-\frac{T_{s2}}{\tau}}}$$

If the sampling rate is $T_s = T_{sn}$, in this case we get

$$H(z) = \frac{K(1-e^{-\frac{T_{sn}}{\tau}})}{z-e^{-\frac{T_{sn}}{\tau}}}$$

Referring to all these conditions, we can say that from a first order continuous in time system we can find many first order discrete-time systems.

Application 3

Convert the discrete transfer function in the following equation into a state space representation.

$$H(z) = \frac{0.5z^2 + 0.5}{z^2 - 1.4z + 0.45}$$

Answer

$$H(z) = \frac{0.5z^2 + 0.5}{z^2 - 1.4z + 0.45} = \frac{0.5\left(z^2 - 1.4z + 0.45\right) + 0.7z + 0.275}{z^2 - 1.4z + 0.45} = 0.5 + \frac{0.7z + 0.275}{z^2 - 1.4z + 0.45}$$

Recall the formula we have derived:

$$\Rightarrow \begin{cases} \begin{bmatrix} x_1(k+1) \\ x_2(k+1) \end{bmatrix} = \begin{bmatrix} 0 & 1 \\ -a_0 & -a_1 \end{bmatrix} \begin{bmatrix} x_1(k) \\ x_2(k) \end{bmatrix} + \begin{bmatrix} b_1 \\ b_0 - a_1 b_1 \end{bmatrix} u(k) \\ y(k) = x_1(k) = \begin{bmatrix} 1 & 0 \end{bmatrix} \begin{bmatrix} x_1(k) \\ x_2(k) \end{bmatrix} \end{cases}$$

$$\Rightarrow \begin{cases} x(k+1) = Ax(k) + Bu(k) \\ y(k) = Cx(k) \end{cases}$$

with

$$A = \begin{bmatrix} 0 & 1 \\ -0.45 & 1.4 \end{bmatrix}, B = \begin{bmatrix} 0.5 \\ 1.2 \end{bmatrix}, C = \begin{bmatrix} 1 & 0 \end{bmatrix}, D = 0.5$$

and

$$C = \begin{bmatrix} C_0 & C_1 & \cdots & C_k \end{bmatrix}$$

The verification process is done by the use of the predefined function **tf2ss** in MatLab:

$$H(z) = C\left(zI - A\right)^{-1} B + D$$

3.10 Conclusions

In this chapter we presented the transfer function of a linear time invariant system. Indeed, the absence or presence of an input sampler is crucial in determining the transfer function of a system. Different cases of the position of the sampler are treated to improve its efficiency and its influence.

3.11 References

Bruce, P. 2017. Essential process control for chemical engineers: bookboon.com ISBN 978-87-403-1655-1.

Franklin, G. and Powell J.D. 1980. Digital control of dynamic systems, Addison Wesley.

Freeman, H. 1965. Discrete-time systems. John Wiley & Sons, New York, USA.

Zhengtao, D. 2013. Nonlinear and adaptive control systems: The Institution of Engineering and Technology.

Haugen, F. 2004. Advanced control, Tapir Academic Press (Norway).

Haugen, F. 2004. Dynamic systems-modeling, analysis and simulation, Tapir Academic Press.

Gang, F. and Rogelio, L. 1999. Adaptive control systems. Butterworth-Heinemann Linacre House, Jordan Hill, Oxford OX2 8DP.

Haugen, F. 2004. PID control, Tapir Academic Press.

John, G.P. and Dimitris, G.M. 2007. digital signal processing: principles, algorithms, and applications, 4th edition.

Arild, W. 2013. Discrete dynamical systems with an introduction to discrete optimization problems: bookboon.com ISBN 978-87-403-0327-8.

Kwakernaak, H. and Sivan, R. 1991. Modern signals and systems. Prentice Hall, New Jersey, USA.

Derek, A. 2015. Control engineering problems with solutions: bookboon.com ISBN 978-87-403-0374-2.

Olsson, G. 2003. Industrial automation, Lund University.

Phillips, C. and Parr, J.M. 1999. Signals, systems, and transforms. Prentice Hall, New Jersey, USA.

Shankar, S. and Marc, B. 1989. Adaptive control: stability, convergence, and robustness: Prentice Hall information and system sciences series.

Rowell, D. and Wormley, D.N. 1997. System dynamics, an introduction. Prentice Hall, New Jersey, USA.

Derek, A. 2014. An introduction to nonlinearity in control systems: bookboon.com ISBN 978-87-7681-790-9.

CHAPTER FOUR

STABILITY OF DISCRETE-TIME SYSTEM

4.1. Introduction

In this chapter, we focus on the stability condition of discrete-time systems in the closed loop, in the first section. In the second section, the global stability definition will be discussed. In the third section, the algebraic stability criterion is detailed. In addition, the stability in the frequency domain is discussed in detail in the fourth section. The root locus is shown in the fifth section. The last section is explaining the system performance.

4.2. Stability definition

The definition of the stability of a discrete-time system is the same as a continuous-time system in that it has the capacity to return to its original position after excitation.

A discrete-time system is given by the transfer function;

$$H(z) = \frac{N(z)}{D(z)} = \frac{N(z)}{\prod_{i=1}^{n}(z - z_i)}$$

A discrete-time system is said to be stable if the denominator of the system, $D(z)$, has no roots or poles outside the unit circle, i.e., a discrete-time system is said to be stable if $|z_i| < 1$.

Let $s_i = \sigma_i + jw_i$, then

$$z_i = e^{s_i T_s} = e^{(\sigma_i + jw_i)T_s} = e^{\sigma_i T_s} e^{jw_i T_s}$$

so

$$|z_i| = \left|e^{s_i T_s}\right| = \left|e^{(\sigma_i + jw_i)T_s}\right| = \left|e^{\sigma_i T_s} e^{jw_i T_s}\right| < 1$$

$$\left|e^{\sigma_i T_s}\right| < 1 \Leftrightarrow \sigma_i < 0$$

where z_i, $i = 1,..,n$, are poles of the denominator $D(z)$ and n is the poles's number.

Marginally stable: a linear time-invariant discrete-time system is marginally stable if the poles are located in the closed unit circle with no repeated poles on the unit circle, i.e., if $|z_i| = 1$ (poles are situated out of the unit circle in the z-plane).

Asymptotically stable: a linear time-invariant discrete-time system is asymptotically stable if its transfer function poles are in the open unit circle. An asymptotically stable system is always Bounded Input - Bounded Output stable.

Unstable: a discrete-time system is said to be unstable if $|z_i| > 1$, i.e., the poles are situated on the unit circle in the z-plane.

In the following figure, the different cases of stability conditions of a system are shown.

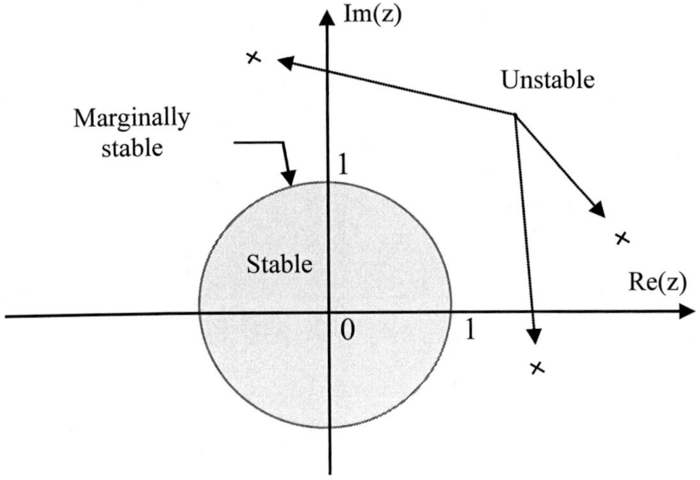

Fig. 4.1. Stability condition in the z-plan

4.2.1. Stability of the 1st order system

Consider a discrete-time 1st order system, in an open loop, given by its difference equation

$$y(k)+a_0y(k-1)=b_0u(k)$$

By applying the z-transform on both sides, we get

$$\left(1+z^{-1}a_0\right)Y(z)=b_0U(z)$$

The transfer function is

$$G(z)=\frac{Y(z)}{U(z)}=\frac{b_0}{1+z^{-1}a_0}=\frac{zb_0}{z+a_0}$$

The unique pole of the transfer function in an open loop is a_0. Consequently, the stability condition is

$$|a_0|<1$$

In the case of a closed loop, it is as given by the following figure.

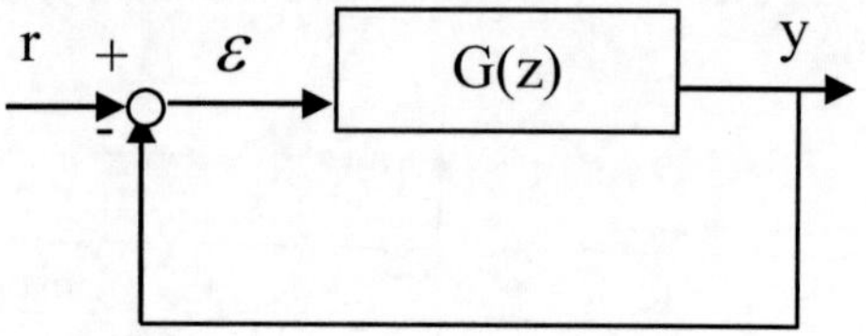

Fig. 4.2. A standard unit feedback discrete-time system

The transfer function in a closed loop is

$$H_{CL}(z)=\frac{G(z)}{1+G(z)}=\frac{\dfrac{b_0}{1+z^{-1}a_0}}{1+\dfrac{b_0}{1+z^{-1}a_0}}=\frac{b_0}{1+z^{-1}a_0+b_0}=\frac{zb_0}{z\left(1+b_0\right)+a_0}$$

The unique pole of the transfer function in a closed loop is $\frac{a_0}{1+b_0}$.

Consequently, the stability condition is

$$\left|\frac{a_0}{1+b_0}\right|<1$$

4.2.2. Stability of the 2nd order system

Consider a discrete-time 2nd order system in an open loop given by its difference equation

$$y(k) + a_1 y(k-1) + a_0 y(k-2) = b_0 u(k)$$

Applying the z-transform on both sides gives:

$$\left(1 + a_1 z^{-1} + a_0 z^{-2}\right) Y(z) = b_0 U(z)$$

The transfer function is:

$$H(z) = \frac{Y(z)}{U(z)} = \frac{b_0}{1 + a_1 z^{-1} + a_0 z^{-2}} = \frac{z^2 b_0}{z^2 + a_1 z + a_0} = \frac{z^2 b_0}{(z - z_1)(z - z_2)}$$

with z_1 and z_2 being the poles of the transfer function.

The two poles of the transfer function, in an open loop, are z_1 and z_2. Consequently, the stability condition is

$$|z_1| < 1 \text{ and } |z_2| < 1$$

The transfer function, in a closed loop case, is

$$H_{CL}(z) = \frac{G(z)}{1 + G(z)} = \frac{\dfrac{b_0 z^2}{z^2 + a_1 z + a_0}}{1 + \dfrac{b_0 z^2}{z^2 + a_1 z + a_0}} = \frac{b_0 z^2}{z^2(1 + b_0) + a_1 z + a_0} = \frac{b_0 z^2}{(z - \alpha_1)(z - \alpha_2)}$$

The two poles of the transfer function, in an open loop, are α_1 and α_2. Consequently, the stability condition is

$$|\alpha_1| < 1 \text{ and } |\alpha_2| < 1$$

As we see, when the order is high or the parameters are variable in time it becomes difficult to find the stability. That is why we study, in the next section, the algebraic stability criterion.

4.3. Algebraic stability criterion

Generally, the roots' computing is difficult due to the high level of the order or the time-varying parameters. Thus, we use an algebraic criterion, without solving the characteristic polynomial, which can be done if the roots are inside the unit circle.

4.3.1. The Jury stability criterion

Let us consider the transfer function

$$H(z)=\frac{N(z)}{D(z)}=\frac{b_0+b_1z+...+b_mz^m}{a_0+a_1z+...+a_nz^n}=\frac{\sum_{i=0}^{m}b_iz^i}{\sum_{i=0}^{n}a_iz^i}$$

with $a_n=1$ and $m\leq n$. Consider the characteristic polynomial $D(z)$ given by

$$D(z)=a_Nz^N+a_{N-1}z^{N-1}+...+a_1z+a_0 \text{ with } a_i\in\mathbf{R}\ i=1,...,n$$

With Jury's test it is easy to check whether all the poles are inside the unit circle, i.e., whether the system is asymptotically stable.

Table 4.1. The table of Jury's criterion

Row	z^0	z^1	z^2	...	z^{n-k}	...	z^{n-3}	z^{n-2}	z^{n-1}
1	a_0	a_1	a_2	...	a_{n-k}	...	a_{n-3}	a_{n-2}	a_{n-1}
2	a_n	a_{n-1}	a_{n-2}	...	a_k	...	a_3	a_2	a_1
3	b_0	b_1	b_2	...	b_{n-k}	...	b_{n-3}	b_{n-2}	b_{n-1}
4	b_{n-1}	b_{n-2}	b_{n-3}	...	b_k	...	b_2	b_1	b_0
5	c_0	c_1	c_2	...	c_{n-k}	...	c_{n-3}	c_{n-2}	0
6	c_{n-2}	c_{n-3}	...	...	c_k	...	c_1	c_0	0
⋮	⋮	⋮			⋮		⋮	0	0
2n-5	p_0	p_1	p_2	p_3	0	0	0	0	0
2n-4	p_3	p_2	p_1	p_0	0	0	0	0	0
2n-3	q_0	q_1	q_2	0	0	0	0	0	0

with:

$$b_j=\begin{vmatrix}a_0 & a_{n-j}\\ a_n & a_j\end{vmatrix}$$

$$c_j=\begin{vmatrix}b_0 & b_{n-1-j}\\ b_{n-1} & b_j\end{vmatrix}$$

$$d_j=\begin{vmatrix}c_0 & c_{n-2-j}\\ c_{n-2} & c_j\end{vmatrix}$$

$$\vdots$$

$$q_0 = \begin{vmatrix} p_0 & p_3 \\ p_3 & p_0 \end{vmatrix}$$

$$q_1 = \begin{vmatrix} p_0 & p_2 \\ p_3 & p_1 \end{vmatrix}$$

$$q_2 = \begin{vmatrix} p_0 & p_1 \\ p_3 & p_2 \end{vmatrix}$$

The necessary and sufficient conditions for the $D(z)$ to have no roots on or outside the unit circle are:

$$D(1) > 0$$

$$\begin{cases} D(-1) > 0 \text{ if } n \text{ is even} \\ D(-1) < 0 \text{ if } n \text{ is odd} \end{cases}$$

$$\left.\begin{array}{l} |a_0| < a_n \\ |b_0| > |b_{n-1}| \\ |c_0| > |c_{n-2}| \\ |d_0| > |d_{n-3}| \\ |q_0| > |q_2| \end{array}\right\} \quad (n-1) \quad \text{constraints}$$

Example: For a second order system, $n = 2$, the Jury's table contains only one row

$$D(1) > 0$$
$$D(-1) > 0$$
$$|a_0| < a_2$$

so

$$D(1) > 0 \Rightarrow a_2 + a_1 + a_0 > 0$$
$$D(-1) > 0 \Rightarrow a_2 - a_1 + a_0 > 0$$
$$-1 < a_0 < 1 \text{ if } a_2 = 1$$

4.3.2. The Routh-Hurwitz stability criterion

This criterion is widely used in the analysis of the stability of the linear time invariant continuous in time system. This criterion allows us to check if the roots of a polynomial are negative. In the case of a discrete-time system, we take $w = \dfrac{z-1}{z+1}$ so the new function is $z = \dfrac{w+1}{1-w}$.

If $|z| < 1$:

$$z = \alpha + j\,\beta \to |z\ | = \sqrt{\alpha^2 + \beta^2} \to |z\ | < 1 \to \alpha^2 + \beta^2 < 1$$

$$w = \frac{\alpha + j\beta - 1}{1 + \alpha + j\beta} = \frac{(\alpha + j\beta - 1).(1 + \alpha - j\beta)}{(1+\alpha)^2 - \beta^2} = \frac{\alpha^2 + \beta^2 - 1 + 2j\beta}{(1+\alpha)^2 - \beta^2}$$

$$\mathrm{Re}(w) = \frac{\alpha^2 + \beta^2 - 1}{(\alpha+1)^2 - \beta^2} < 0$$

If $w = a + jb$ $\quad with \quad a < 0$ so

$$z = \frac{1 + a + jb}{1 - a - jb} = \frac{1+w}{1-w}$$

$$|z| = \left|\frac{w+1}{1-w}\right| = \left|\frac{a + jb + 1}{1 - a - jb}\right| < 1$$

$$(1+a)^2 + b^2 < (1-a)^2 + b^2 \Rightarrow (1+a)^2 < (1-a)^2$$

$$(1+a)^2 - (1-a)^2 < 0 \Rightarrow 2(2a) < 0 \to 4a < 0$$

Using the discrete-time transfer function $H(z)$:

$$H(z) = \frac{N(z)}{D(z)}$$

with $D(z)$ being the characteristic polynomial

$$D(z) = a_N z^N + a_{N-1} z^{N-1} + \ldots + a_1 z + a_0 \text{ with } a_i \in \mathbf{R}\ i = 1, \ldots, n$$

Where $z = \dfrac{1+w}{1-w}$, we find

$$D(z=\frac{1+w}{1-w})=a_0+a_1\frac{1+w}{1-w}+...+a_{n-1}\frac{(1+w)^{n-1}}{(1-w)^{n-1}}+a_n\frac{(1+w)^n}{(1-w)^n}$$

$$=a_0\frac{(1-w)^n}{(1-w)^n}+a_1\frac{(1+w)(1-w)^{n-1}}{(1-w)^n}+...+a_{n-1}\frac{(1+w)^{n-1}(1-w)}{(1-w)^n}+a_n\frac{(1+w)^n}{(1-w)^n}$$

$$(1-w)^n D(\frac{1+w}{1-w})=a_0(1-w)^n+a_1(1+w)(1-w)^{n-1}+...+a_{n-1}(1+w)^{n-1}(1-w)+a_n(1+w)^n$$

$$(1-w)^n D(\frac{1+w}{1-w})=\alpha_0+\alpha_1 w+...+\alpha_{n-1}w^{n-1}+\alpha_n w^n$$

The necessary and sufficient condition for this criterion is that all roots of the first column have the same sign.

Table 4.2. The table of Routh's criterion

Row 1	α_n	α_{n-2}	α_{n-4}	...	α_0
Row 2	α_{n-1}	α_{n-3}	α_{n-5}		α_1
Row 3	β_1	β_2	β_3	...	β_n
Row 4	δ_1	δ_2	δ_3	...	δ_n
...					

with

$$\beta_1=\frac{\alpha_{n-1}\alpha_{n-2}-\alpha_n\alpha_{n-3}}{\alpha_{n-1}}$$

$$\beta_2=\frac{\alpha_{n-1}\alpha_{n-4}-\alpha_n\alpha_{n-5}}{\alpha_{n-1}}$$

$$\delta_1=\frac{\beta_1\alpha_{n-3}-\alpha_{n-1}\beta_2}{\beta_1}$$

...

Example: Study the stability of the closed loop of the system using the Routh criterion given by the following figure.

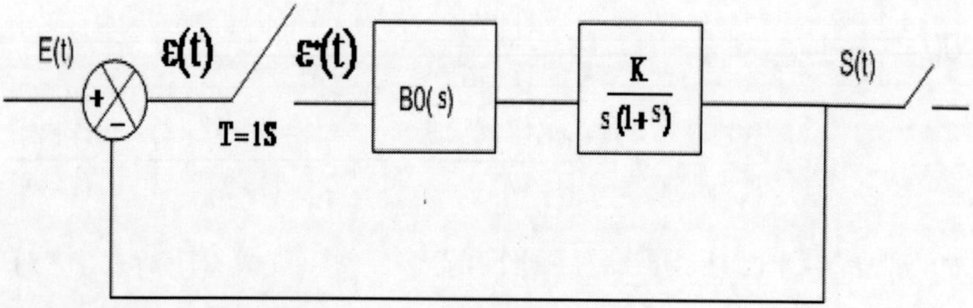

Fig. 4.3. Unit feedback discrete-time second order system

The transfer function in the open loop is

$$H_{OL}(z) = Z\left[B_0(s) \frac{K}{s(s+1)} \right]$$

$$= K\frac{0.37z + 0.26}{(z-1)(z-0.37)}$$

The transfer function in the closed loop is

$$H_{CL}(z) = \frac{0.37Kz + 0.26K}{z^2 - 1.37z + 0.37 + 0.37Kz + 0.26K}$$

$$= \frac{K(0.37z + 0.26)}{z^2 + (-1.37 + 0.37K)z + 0.26K + 0.37}$$

$$= \frac{N(z)}{D(z)}$$

with $D(z) = z^2 + (-1.37 + 0.37K)z + 0.26K + 0.37$

We apply the transform of w:

$$D(z = \frac{1+w}{1-w}) = (\frac{1+w}{1-w})^2 + (-1.37 + 0.37K)(\frac{1+w}{1-w}) + 0.26K + 0.37$$

$$(1-w)^2 D(z = \frac{1+w}{1-w}) = (1+w)^2 + (-1.37 + 0.37K)(1+w)(1-w) + (0.26K + 0.37)(1-w)^2$$

$$D(w) = (1+w)^2 + (-1.37+0.37K)(1-w^2) + (0.26K+0.37)(1-w^2)$$
$$= (2.74-0.11K)w^2 + (1.26-0.52K)w + 0.63K$$
$$= a_2 w^2 + a_1 w + a_0$$

with

$$\begin{cases} a_2 = 2.74 - 0.11K \\ a_1 = 1.26 - 0.52K \\ a_0 = 0.63K \end{cases}$$

K is the static gain which is positive.

$$1.26 - 0.52K > 0 \rightarrow K < \frac{1.26}{0.52} = 2.42$$

$$2.74 - 0.11K > 0 \rightarrow K < \frac{2.74}{0.11} = 24$$

The condition of stability of the closed loop of the system using the Routh criterion is:

$$0<K<2.42$$

The lack of an algebraic criterion demands the use of another method which is based on the domain frquency.

4.4. Domain frequency stability

4.4.1. Bode diagrams for continuous-time systems

The Bode diagram is the frequency responses (both magnitude response and phase response with respect to the frequency).

Consider a standard unit feedback continuous-time system:

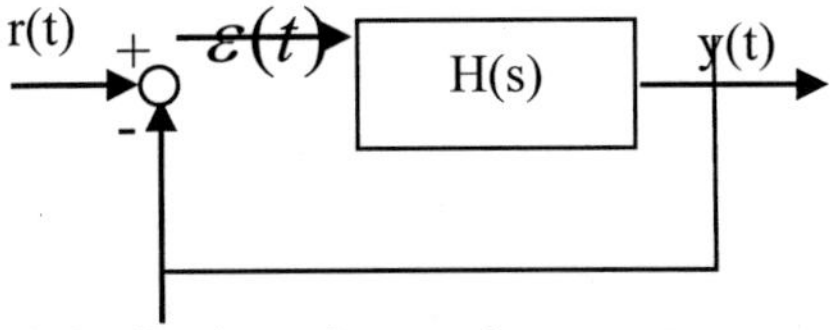

Fig. 4.4. A standard unit feedback continuous-time system

The transfer function in the open loop (OL) is

$$H_{OL}(s) = \frac{N_{OL}(s)}{D_{OL}(s)}$$

and the closed-loop (CL) transfer function $H_{CL}(s)$ is

$$H_{CL}(s) = \frac{H_{OL}(s)}{1 + H_{OL}(s)} = \frac{N_{CL}(s)}{D_{CL}(s)}$$

Interestingly, if the open-loop transfer function $H_{OL}(s)$ is stable, then its frequency responses can be easily used to determine the stability of the closed-loop system $H_{CL}(s)$. As such, it is often that we plot the Bode diagram for the open-loop system $H_{OL}(s)$.

Magnitude and phase responses are defined as:

Magnitude response: $|H_{OL}(jw)|$

Phase response: $\arg H_{OL}(jw)$

Example: The Bode diagrams of the open-loop system of a given continuous in time system function $H_{OL}(s) = \dfrac{1}{s^2 + s + 1}$ are given by the use of a predefined MatLab function.

The continuous-time transfer function:

```
>> Hs=tf(1,[1 1 1])
Hs =
        1
  ----------
  s^2 + s + 1
 Continuous-time transfer function.
```

The Bode presentation of continuous-time system

```
>> w=logspace(-2,1,1000)';
>> bode(Hs,w,'b'),grid%continuous
```

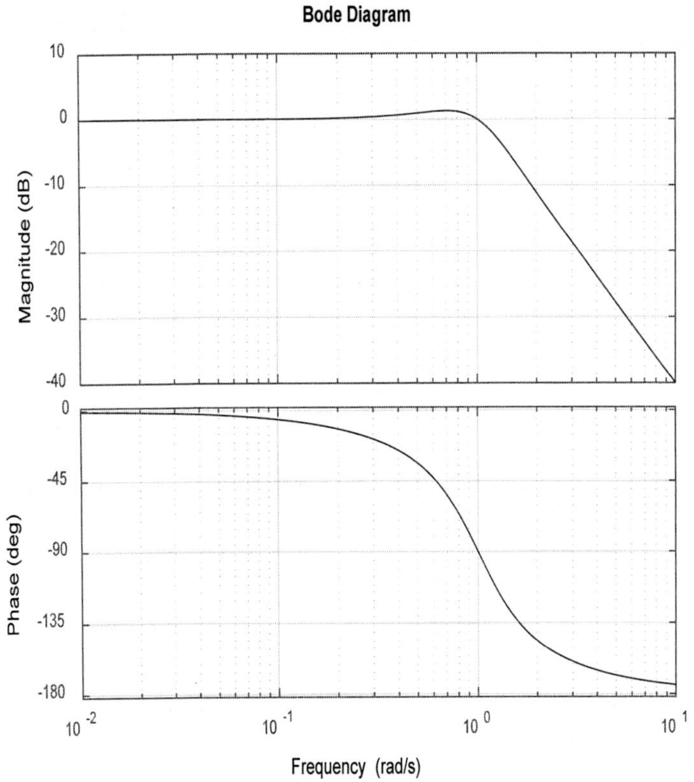

Fig. 4.5. Bode diagram of continuous-time system

The gain and phase margins in the Bode diagram of a continuous-time system are presented in the following figure.

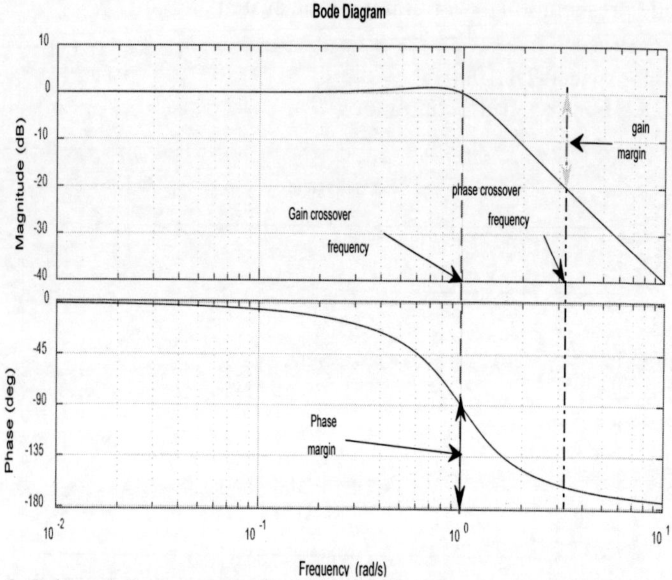

Fig. 4.6. Gain and phase margins in the Bode diagram of a continuous-time system

4.4.2. Nyquist stability theory in continuous-time systems

The Nyquist plot draws the frequency response of the open-loop system on the same complex plane instead of separating the magnitude and phase responses into two different diagrams as in the Bode plot.
The Nyquist presentation of a continuous-time system:

```
>> w=logspace(-2,1,1000)';
>> Nyquist (Hs,w,'b'),grid%continuous
```

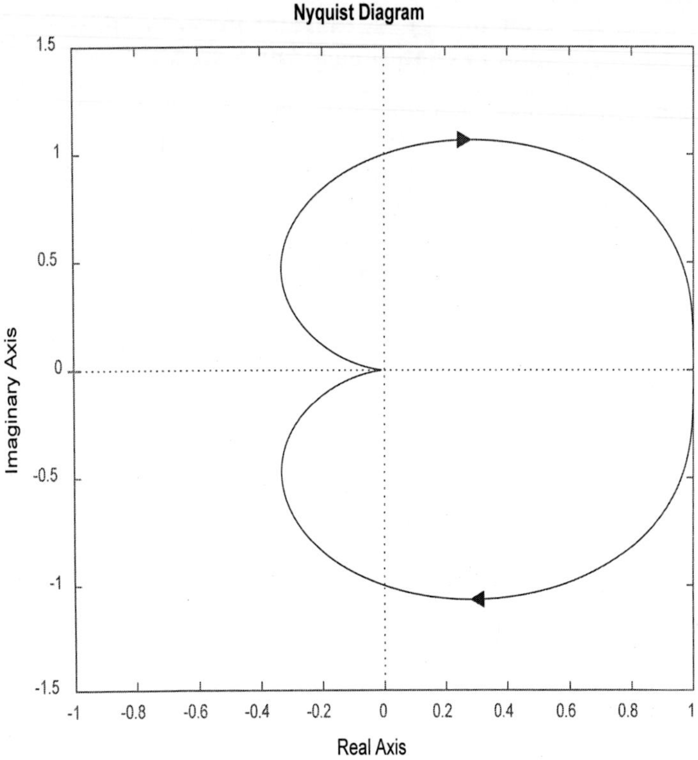

Fig. 4.7. Nyquist diagram of a continuous-time system

Nyquist Stability Criterion:

Let n_c be the number of clockwise encirclements of the point $\{-1\}$ in the Nyquist plot, and n_p be the number of unstable poles of the open-loop system $\mathrm{H_{OL}(s)}$. Then, the number of unstable poles of the closed-loop system $\mathrm{H_{CL}(s)}$, denoted by n_z, is given by $n_z = n_p + n_c$. If $\mathrm{H_{OL}(s)}$ is stable, i.e., $n_p = 0$, then n_c has to be zero in order to guarantee the stability of the closed-loop system $\mathrm{H_{CL}(s)}$.

4.4.3. Gain and phase margins in the Nyquist plot in continuous-time systems

Assuming that the open-loop system is stable, the gain margin (GM) and phase margin (PM) can be found from the Nyquist plot by zooming into the region in the neighborhood of the origin.

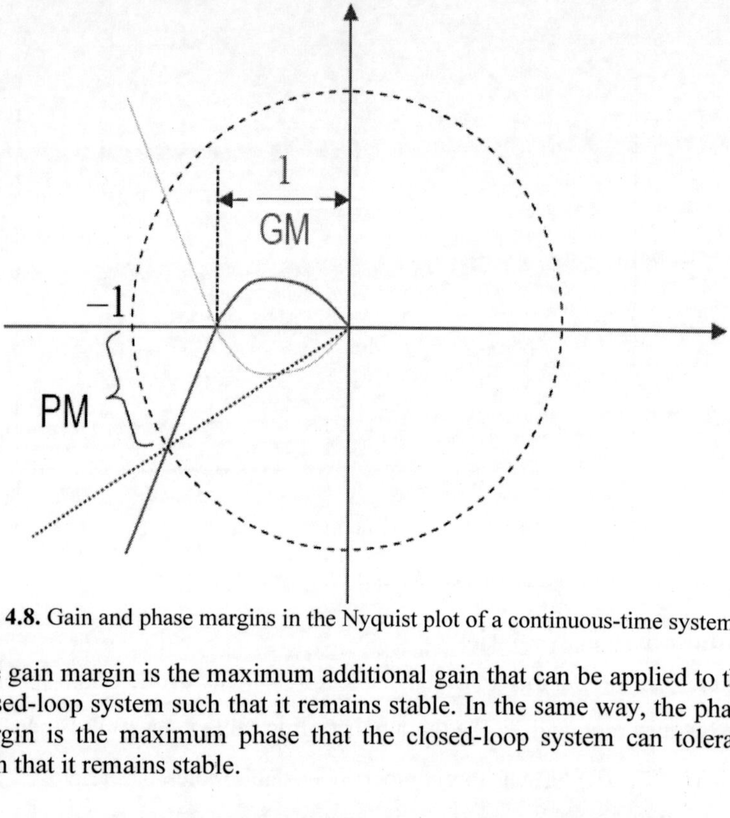

Fig. 4.8. Gain and phase margins in the Nyquist plot of a continuous-time system

The gain margin is the maximum additional gain that can be applied to the closed-loop system such that it remains stable. In the same way, the phase margin is the maximum phase that the closed-loop system can tolerate such that it remains stable.

4.4.4. Bode diagrams for discrete-time systems

For a continuous-time system $H(s)$, the frequency response is $H(jw)$, $w \in [0, +\infty[$. It can be graphically presented in the complex plane as the Nyquist curve or as amplitude/phase curves as a function of frequency (Bode diagram).

Correspondingly, the frequency response of a discrete-time system $H(z)$ is $H(e^{jwT_s}), wT_s \in [0, \pi[$. This can also be presented graphically as a discrete Nyquist or discrete Bode diagram. The difference is, that in the discrete case only the frequency interval $wT_s \in [-\pi, \pi[$ is considered.

The Bode diagram for discrete-time systems consists of both the magnitude and phase responses of a discrete-time system, which are defined as follows:

$$\text{Magnitude response: } \left|H_{OL}(z)\right|_{z=e^{jwT_s}}$$

$$\text{Phase response: } \arg H_{OL}(z)_{z=e^{jwT_s}} \qquad -\pi \leq wT_s \leq \pi$$

Consider a standard unit feedback discrete-time system:

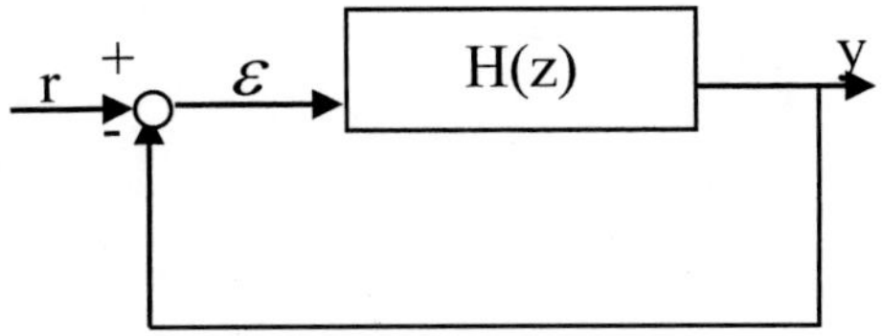

Fig. 4.9. A standard unit feedback discrete-time system

Example: Let us draw the Bode diagram for the open-loop system which is previously defined with the sampling period $T_s = 1s$.

The discrete in time transfer function:

```
>>Hz=c2d(Hs,1)
Hz =
    0.3403 z + 0.2417
  -----------------------
  z^2 - 0.7859 z + 0.3679
Sample time: 1 second
Discrete-time transfer function.
```

```
>> w=logspace(-2,1,1000)';
>> bode(Hz,w,'b'),grid
```

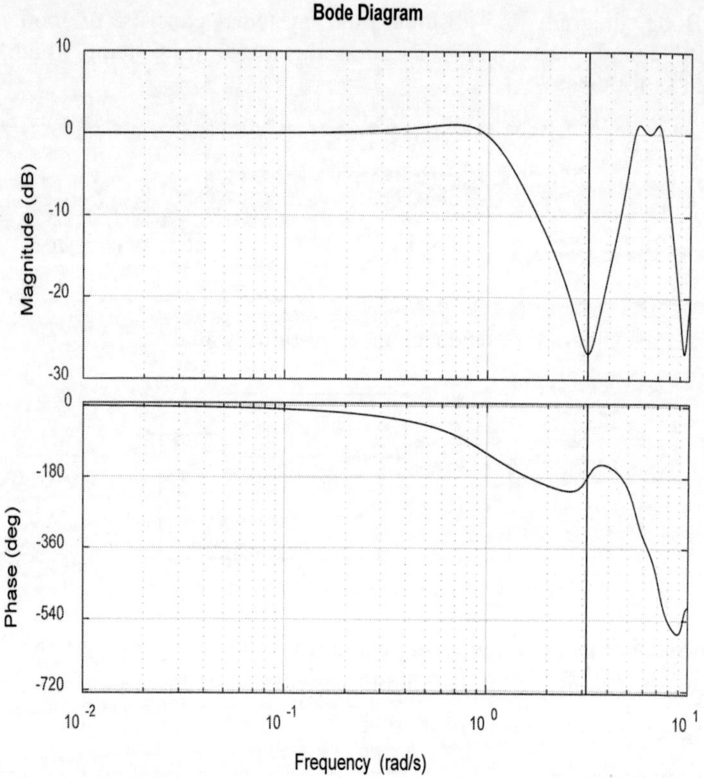

Fig. 4.10. Bode diagram of discrete-time system

The gain and phase margins in the Bode diagram of a discrete-time system are presented in the following figure.

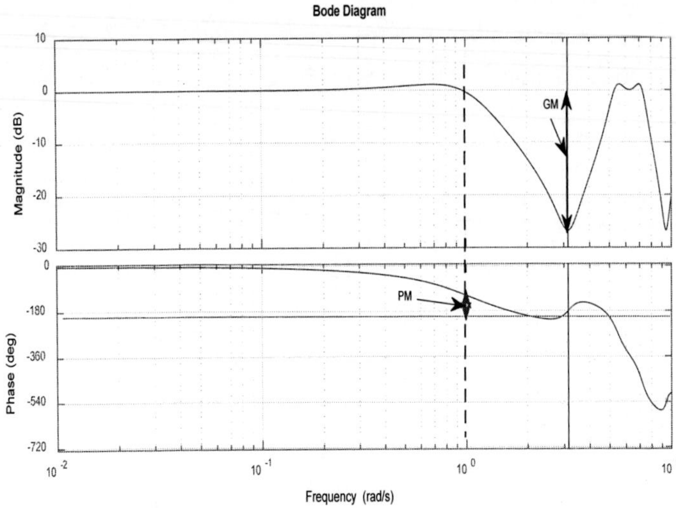

Fig. 4.11. Gain and phase margins in the Bode diagram of a discerete-time system

Note: Both magnitude and phase responses repeat after 2π/T.

4.4.5. Discrete Nyquist stability criterion

The transfer function in the open loop (OL) is

$$\mathrm{H_{OL}(z) = G(z) = \frac{N_{OL}(z)}{D_{OL}(z)}}$$

The transfer function in the closed loop (CL) is

$$\mathrm{H_{CL}(z) = \frac{H_{OL}(z)}{1 + H_{OL}(z)} = \frac{N_{CL}(z)}{D_{CL}(z)}}$$

So, the characteristic equation is

$$D_{CL}(z) = 1 + H_{OL}(z) = 0$$

The stability can be determined by using the open loop $H_{OL}(z)$ Nyquist diagram. $H_{OL}(e^{jwT_s})$ (the open loop Nyquist curve) encircles the point $\{-1\}$ n_c times clockwise.

$$n_c = n_z - n_p$$

in which n_z is the number of zeros and n_p is the number of poles of the characteristic equation outside the unit circle.

This fact can be applied in stability analysis. The characteristic equation has the form

$$
\begin{aligned}
D_{CL}(z) &= 1 + H_{OL}(z) \\
&= 1 + \frac{N_{OL}(z)}{D_{OL}(z)} \\
&= \frac{D_{OL}(z) + N_{OL}(z)}{D_{OL}(z)} \\
&= \frac{N_{CE}(z)}{D_{CE}(z)}
\end{aligned}
$$

The open loop poles are the same as the poles of the characteristic equation. The zeros of the characteristic equation determine stability so that if the characteristic equation has zeros outside the unit circle, the closed loop system is unstable. The stability criterion is thus obtained by setting $n_z = 0$ and by demanding that the Nyquist curve encircles the point $\{-1\}$ n_p times anti-clockwise.

$$n_z = n_c + n_p = 0$$

The criterion becomes simple, if the open loop pulse transfer function has no poles outside the unit circle. Then the Nyquist curve should not encircle the point $\{-1\}$ at all.

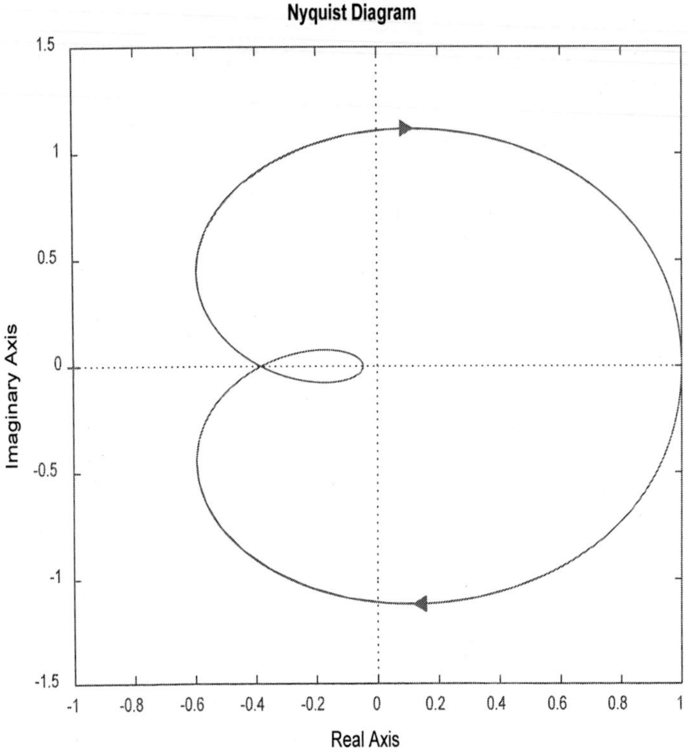

Fig. 4.12. Nyquist diagram of discrete-time system

Let us use the MatLab function to compare these two systems, the Bode presentation of continuous-time and the discrete-time system:

```
>> w=logspace(-2,1,1000)';
>> bode(Hs,w,'k')%continuous
   hold
   bode(Hz,w,'b')%discrete
```

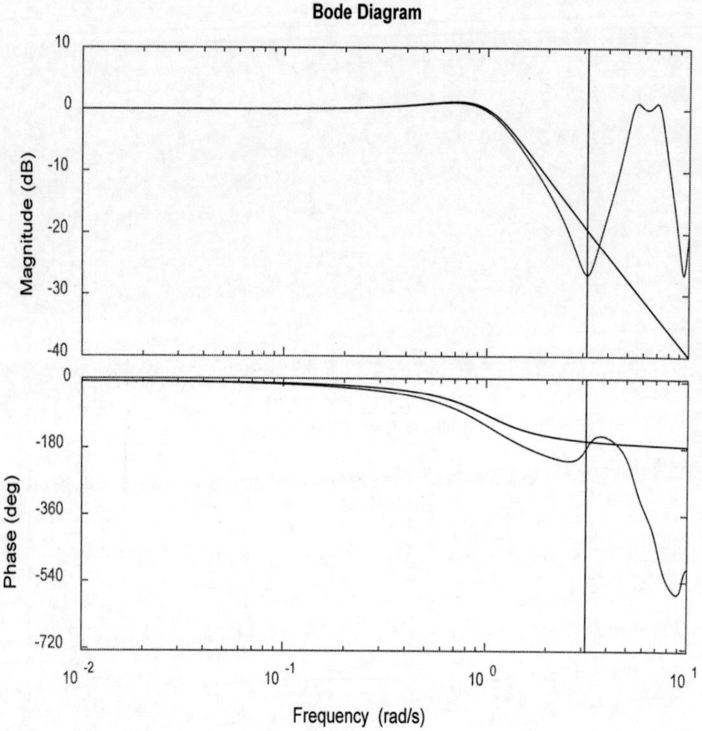

Fig. 4.13. The Bode presentation of continuous-time and discrete-time systems

The Nyquist diagram of continuous-time and discrete-time systems is

```
>> nyquist(Hs,w,'k')
   hold
   nyquist(Hz,w,'r')
```

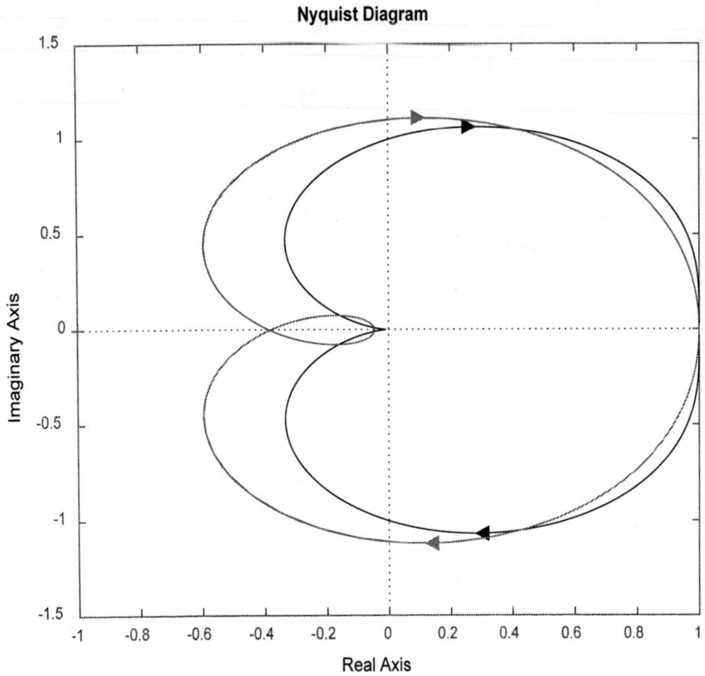

Fig. 4.14. The Nyquist presentation of continuous-time and discrete-time systems

Example:

A process is controlled with a discrete proportional controller, which has the gain $K = 1$.

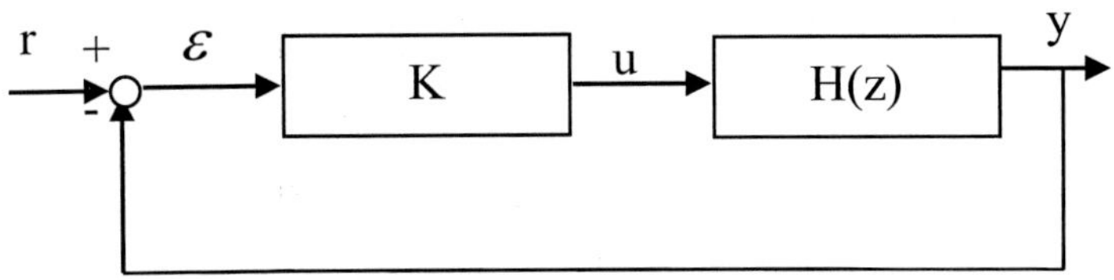

Fig. 4.15. A process is controlled with a discrete proportional controller

where

$$H(z) = \frac{0.4}{(z-0.5)(z-0.2)}$$

The discrete Nyquist diagram is constructed with MatLab:

```
>> Hz=zpk([],[0.2 0.5],0.4,1);
Sample time: 1 seconds;
Discrete-time zero/pole/gain model.
>> nyquist(Hz)
```

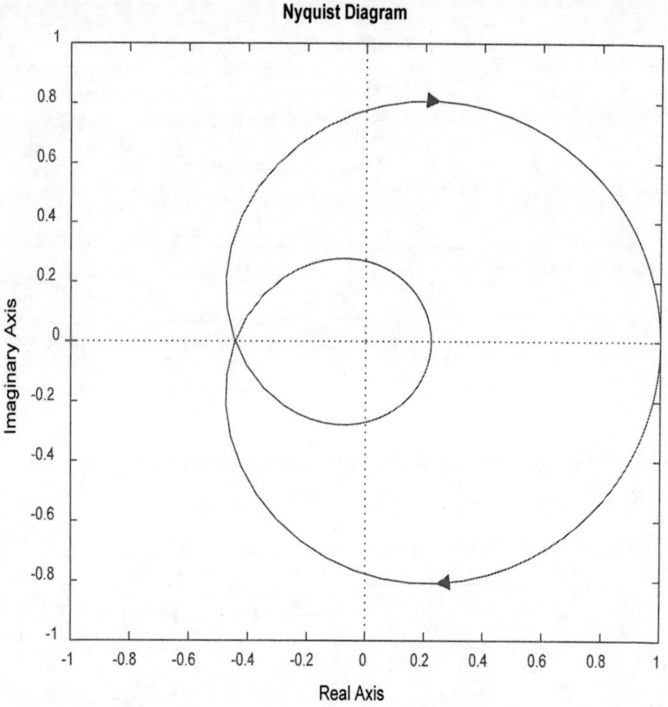

Fig. 4.16. The Nyquist diagram of a process that is controlled with a discrete proportional controller

One way to find the interception point in the real axis is to use the *zoom*-command. By inspection, the point is approximately -0.4416.

The magnitude can thus be multiplied with (1/0.4416) to reach the critical point $\{-1\}$. The controlled system is stable when $K < \frac{1}{0.4416} = 2.26$

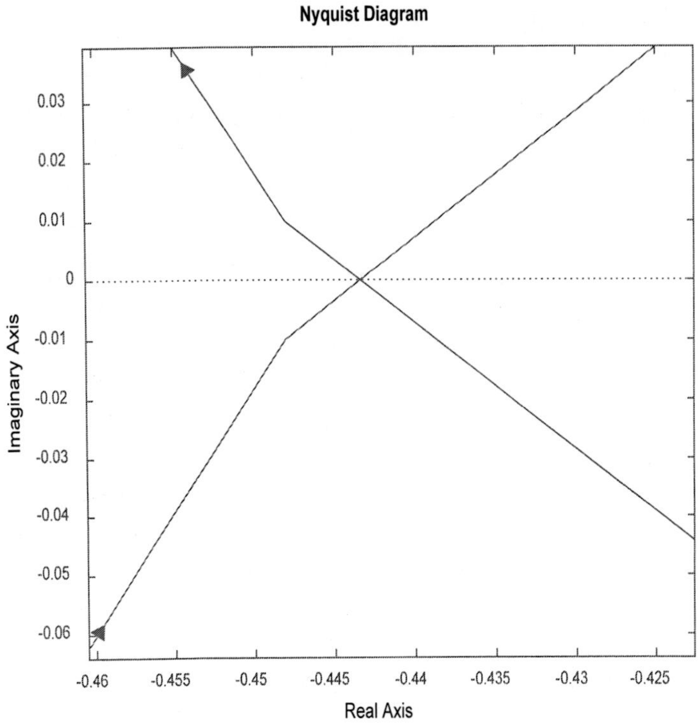

Fig. 4.17. Zoom of a Nyquist diagram of a process that is controlled with a discrete proportional controller

Stability can also be determined by direct calculus from the pulse transfer function:

$$H(z) = \frac{0.4}{(z-0.5)(z-0.2)}$$

Substitute z with $e^{jwT_s} = e^{jw} = \cos w + j \sin w$, (Euler formula), which gives the frequency response $H(e^{jwT_s})$, $T_s = 1s$.

$$H(z)=\frac{0.4}{(e^{jw}-0.5)(e^{jw}-0.2)}$$

$$=\frac{0.4}{(\cos w+j\sin w-0.5)(\cos w+j\sin w-0.2)}$$

$$=\frac{0.4}{(\cos^2 w-\sin^2 w-0.7\cos w+0.1)+j(2\sin w\cos w-0.7\sin w)}$$

$$=\frac{0.4}{(2\cos^2 w-0.7\cos w-0.9)+j(2\sin w\cos w-0.7\sin w)}$$

Setting the imaginary part to 0, the interception point with the real axis is obtained as:

$$2\sin w\cos w-0.7\sin w=0\Rightarrow \sin w(0.7-2\cos w)=0$$

$$\Rightarrow(\sin w=0)\vee(0.7-2\cos w=0)$$

$$\Rightarrow(\sin w=0)\vee\left(\cos w=\frac{7}{20}\right)$$

$$\Rightarrow(w=0)\vee\left(w=ar\cos\frac{7}{20}\right)$$

The frequency 0 describes the starting point in the Nyquist curve and the frequency *arcos*(7/20) is the interception point with the real axis. Substitute it into the frequency response function

$$H\left(e^{j\left(ar\cos(\frac{7}{20})\right)}\right)=\frac{0.4}{2(\frac{7}{20})^2-0.7(\frac{7}{20})-0.9}=-0.444$$

The interception point is -0.444. The gain of the controller K can be multiplied by the factor (1/0.444) in order that it crosses it at point –1.

The controlled system is stable when $K<\frac{9}{4}=2.25$

The symbolic frequency response is calculated with the MatLab symbolic toolbox

```
>> Hz='0.4/((z-0.5)*(z-0.2))'
Hz =
0.4/((z-0.5)*(z-0.2))
>> z='cos(w)+i*sin(w)'
z =
cos(w)+i*sin(w)
>>h1 =subs(Hz,z)
h1 =
 0.4/((cos(w) + i*sin(w) - 0.5)*(cos(w) + i*sin(w) - 0.2))
>> h2=simplify(h1)
h2 =
 0.4/((cos(w) + i*sin(w) - 0.5)*(cos(w) + i*sin(w) - 0.2))
```

4.5. Roots locus method

4.5.1. Position of poles and zeros of system in s-domain in open loop

The poles of a system are represented by the symbol $\times$, these poles are the values that cancel the denominator of the transfer function. However, the zeros of a system are represented by o which are the values that cancel the numerator of the transfer function.

Example: let us consider a continuous-time of first order system given by the following transfer function.

$$H(s) = \frac{K}{1+\tau s}$$

According to this transfer function, this system has a unique pole $-\frac{1}{\tau}$.

The further this pole is from the origin, the faster the system is:

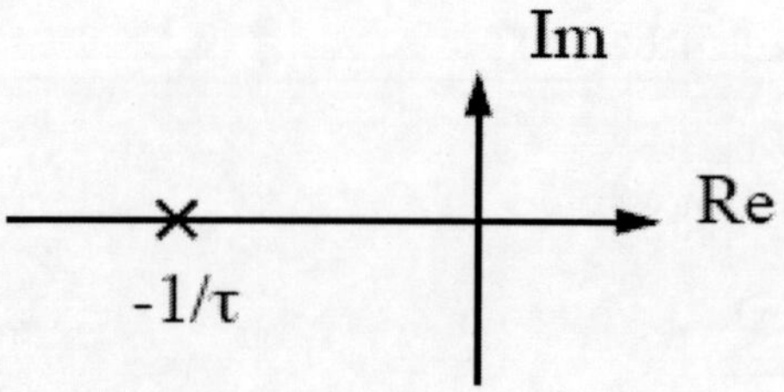

Fig. 4.18. Roots locus of first order system

Example: let us consider a continuous-time of second order system given by the following transfer function.

$$H(s) = \frac{w_n^2}{s^2 + 2\xi w_n s + w_n^2}$$

The discriminant of the denominator is

$$\Delta = (2w_n)^2 (\xi^2 - 1)$$

If $\Delta > 1$, there are two real poles. The time constant is related to the position of the closed pole to the origin (dominant pole).

If $\Delta = 0$, there is a double pole in w_n.

If $\Delta < 1$, there are two complex conjugate poles.

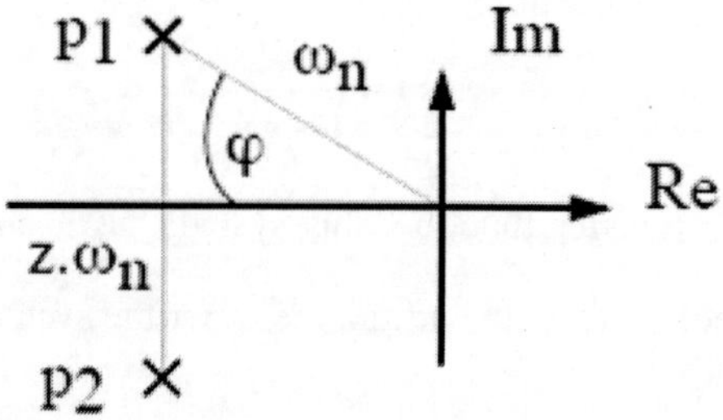

Fig. 4.19. Roots locus of second order system

where

$$\tan g\varphi = \frac{\sqrt{1 - \Delta^2}}{\Delta}$$

4.5.2. Position of poles and zeros of system in s-domain in closed loop

We consider a system in closed-loop (CL) with unitary return such as the transfer function in open-loop (OL) is given by:

$$H_{OL}(s) = K\frac{N_{OL}(s)}{D_{OL}(s)}$$

where $N_{OL}(s)$ and $D_{OL}(s)$ are polynomials (respectively numerator and denominator of $H_{OL}(s)$) and K is the gain of the system. This system is equivalent to a transfer function $H_{CL}(s)$:

$$H_{CL}(s) = \frac{H_{OL}(s)}{1 + H_{OL}(s)} = \frac{KN_{OL}(s)}{D_{OL}(s) + KN_{OL}(s)} = \frac{N_{CL}(s)}{D_{CL}(s)}$$

The poles of this system in CL verify the following characteristic equation:

$$D_{CL}(s) = D_{OL}(s) + KN_{OL}(s)$$

The position of the poles in CL depend on K. If this factor, K, is an adjustable variable of the system, the place of the Evans locus or one of poles is the geometric place of the roots of the characteristic equation plotted in the complex plane when K varies from 0 to ∞.

Knowing this place makes it possible to predict the behavior of the system in CL when K varies because the position of the poles provides more information than the speed and the stability of the system.

4.5.3 Property and construction rules in s-domain

Number of branches: the number of poles in CL is equal to the number of poles in OL, and it is the order of the system.

Symmetry about the real axis: whatever the value of K, the complex poles always go in conjugate pairs.

Starting point: for $K \to 0$, the characteristic equation becomes $D_{CL}(s) = 0$ and we find the poles in OL.

Arrival point: for $K \to \infty$, the characteristic equation becomes $N_{CL}(s) = 0$ and we find the zeros of the transfer function in OL, thus they are the end points of certain branches.

Endless branches: the branches that do not go to an end point, but go on forever. If n is the number of poles and m is the number of zeros of the OL system, then the asymptotes are characterised by the odd multiples of $\frac{\pi}{n-m}$ as asymptotic directions and $\frac{1}{n-m}\left(\sum poles - \sum zeros\right)$ as the abscissa of these asymptotes on the axis of real numbers.

Position of the place belonging to the real axis: a point M of the axis of the real numbers belongs to the place if and only if the number of poles and real zeros located to the right of M is odd.

Connection point: they are the points where the place leaves or joins the real axis. This corresponds to values of K such that the system in CL has double poles.

- we are looking for solutions to the equation:

$$\sum_{i=1}^{n} \frac{1}{s - s_i} = \sum_{j=1}^{m} \frac{1}{z - z_j}$$

where s_i and z_i are respectively the poles and zeros of the transfer function in OL, n is the number of poles (order) and m the number of zeros of the system in OL.

- we set $y(x) = \frac{D(x)}{N(x)}$ and we look for the values of x which cancel $\frac{dy}{dx}$.

Intersection with the imaginary axis: if the place intersects with the imaginary axis, it is because, for certain values of K, the transfer function in CL has pure imaginary poles. To find the value of y and K, we set $s = jy$ and then separate the real part and the imaginary part from the characteristic equation $D_{CL}(s)$.

Tangent at a starting or ending point: if this point is real, the tangent is horizontal except if it is about a point of separation, in which case, the tangent is vertical.

The tangent at the start of a complex pole is given by:

$$\theta_s = \pi + \sum \alpha_i - \sum \beta_j$$

The tangent at the end point of a complex zero is given by:

$$\theta_e = \pi + \sum \beta_j - \sum \alpha_i$$

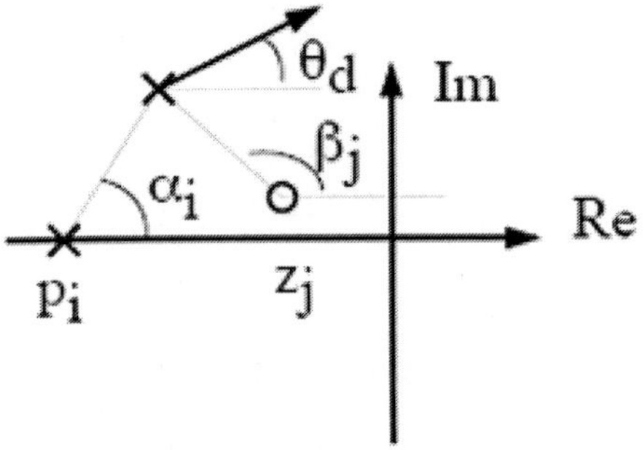

Fig. 4.20. Construction of a tangent at a start or end point

4.5.4. Positions of poles and zeros of a system in a z-domain in a closed loop

As given in the case of the continuous-time system, the stability of the control system depends on the distribution of poles in the closed loop in the z-plane. Thus, when we study the roots locus we should look for poles of the transfer function in the closed loop whether or not a variable gain term, K, varies from 0 to $+\infty$ in the graphic windows (z-plane).
The discrete transfer function in the open loop, is

$$H_{OL}(z) = K\frac{N_{OL}(z)}{D_{OL}(z)}$$

In the canonical, form, it is given by:

$$H_{OL}(z) = K \frac{\prod_{j=1}^{m}(z - z_j)}{\prod_{i=1}^{n}(z - p_i)}$$

Where z_j, $j = 1, \ldots, m$, and p_i , $i = 1, \ldots, n$, are respectively zeros and poles of the transfer function.

In the closed loop the transfer function is

$$H_{CL}(z) = \frac{H_{OL}(z)}{1 + H_{OL}(z)}$$

$$= \frac{K \frac{N_{OL}(z)}{D_{OL}(z)}}{1 + K \frac{N_{OL}(z)}{D_{OL}(z)}}$$

$$H_{CL}(z) = \frac{KN_{OL}(z)}{D_{OL}(z) + KN_{OL}(z)} = \frac{N_{CL}(z)}{D_{CL}(z)}$$

Using the canonical form of the transfer function in the open loop case, the stability condition depends on the characteristic equation

$$D_{CL}(z) = D_{OL}(z) + KN_{OL}(z)$$

$$D_{CL}(z) = \prod_{i=1}^{n}(z - p_i) + K \prod_{j=1}^{m}(z - z_j)$$

$$= (z - p_1)(z - p_2)\ldots(z - p_n) + K\,(z - z_1)(z - z_2)\ldots(z - z_m)$$

$$= 0$$

Any point on the root locus must satisfy the magnitude condition:

$$K \frac{|z - z_1||z - z_1|\ldots|z - z_m|}{|z - p_1||z - p_1|\ldots|z - p_n|} = 1$$

The angle conditions are:

$$\arg(z - z_1) + \ldots + \arg(z - z_m) - \arg(z - p_1) - \ldots - \arg(z - p_n) = i\pi$$

with $i = \cdots -3, -11, 3, 5 \cdots$

The angle condition is used to locate points on the root locus and the magnitude condition gives the value of K at that point. From the characteristic equation, in the z-plane, if the variable gain term K varies in $[0,+\infty[$ then all representative points of the poles of the closed loop draw a curved E called the roots locus or EVANS locus.

4.5.5. Property and construction rules in a z-domain

- The number of loci is equal to the order n of the characteristic equation of the transfer function in the closed loop $H_{CL}(z)$.
- The loci start (i.e. $K=0$) at the n poles of the transfer function $H_{OL}(z)$ in the open loop.
- The root loci end (i.e. $K=+\infty$) at the m zeros of the transfer function $H_{OL}(z)$, and if $m<n$ then the remaining $(n-m)$ loci tend to infinity.
- Portions of the real axis are sections of a root locus if the number of poles and zeros lying on the axis to the right is odd.
- Those terminating loci at infinity tend towards asymptotes at relative angles to the positive real axis given by:

$$\frac{\pi}{n-m},\frac{3\pi}{n-m},\frac{5\pi}{n-m},\ldots,\frac{\left(2(n-m)-1\right)\pi}{n-m}$$

- The intersection of the asymptotes on the real axis occurs at the 'gravity centre' of the pole – the zero configuration of $H_{OL}(z)$:

$$z=\frac{1}{n-m}\left(\sum poles\ of\ H_{OL}(z)-\sum zeros\ of\ H_{OL}(z)\right)$$

- The intersection of the root - loci with the unit circle can be calculated by the use of Jury, Routh or some other geometrical analysis (but on some plots).
- The breakaway points (points at which multiple roots of the characteristic polynomial occur) of the root locus are the solutions of $\frac{dK}{dz}=0$ (not all the solutions are necessarily breakaway points).

Example 1

If the first order transfer function is $H_{OL}(z) = \dfrac{1}{z-1}$, then the root locus is

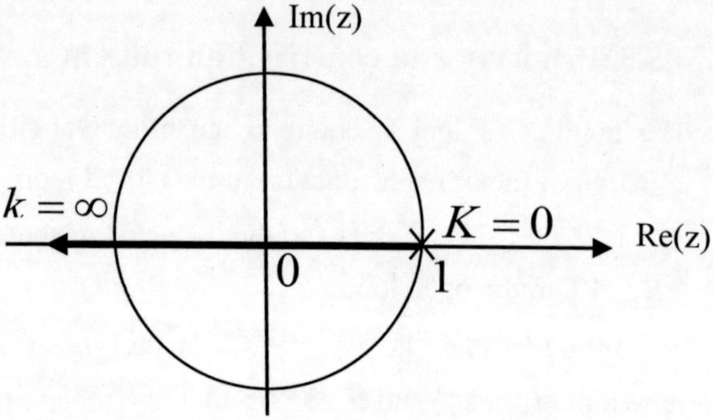

Fig. 4.21. Case of the root locus of the first order transfer function

In another case, if the transfer function is $H_{OL}(z) = \dfrac{z}{z-1}$, then the root locus is

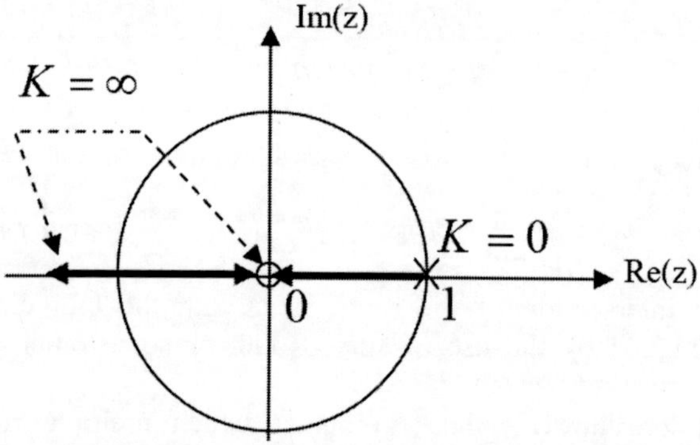

Fig. 4.22. Case of the root locus of the first order transfer function

Example 2

If the second order transfer function $H_{OL}(z) = \dfrac{z}{(z-1)(z-2)}$, then the root locus is

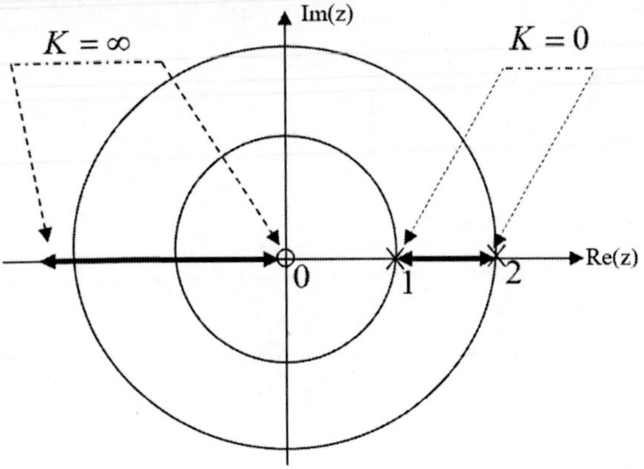

Fig. 4.23. Case of the root locus of the second order transfer function

If the transfer function $H_{OL}(z)=\dfrac{z}{(z-p_1)(z-\overline{p}_1)}$, and $p_1<1$, then the root locus is

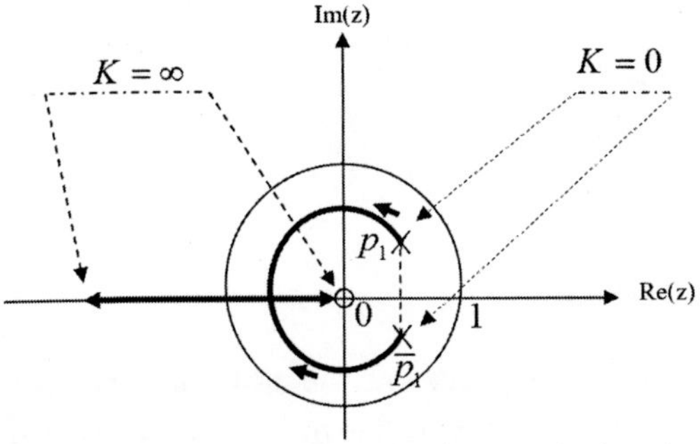

Fig. 4.24. Case of the root locus of the second order transfer function

If the transfer function $H_{OL}(z)=\dfrac{(z-z_1)(z-z_2)}{(z-p_1)(z-\overline{p}_1)}$, $p_1<1$, $z_1<1$, then the root locus is

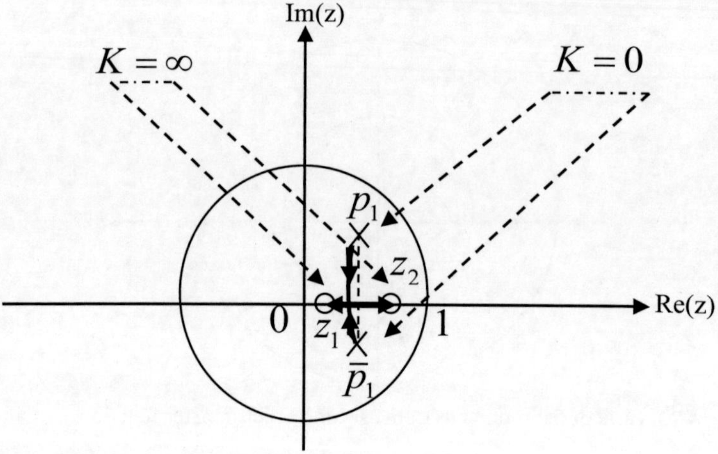

Fig. 4.25. Case of the root locus of the second order transfer function

Example 3

The transfer function of a discrete-time system $H(z)=\dfrac{0.26z+0.3}{(z-1)(z-0.25)}$ corresponding to a continuous time transfer function $H(s)$ is $\dfrac{1}{s(s+2)}$ with $T_s=1s$.

--1-- Manual method:

In this case $m=1$, so we have one point to start, and $n=2$, so we have two points to finish - however $n-m=1$ so we have no asymptotical direction.

With $K=0.2k$ and k as a variable gain term in the graphic windows, this is also a regulation parameter:

If $k=0$, then $z=p_i$ and the n poles of the open loop $H_{OL}(z)$ are the starting points of E.

If $k=\infty$, then $z=z_i$ and the m zeros of the open loop $H_{OL}(z)$ are the finishing points of E.

From this assumption, there are n starting points and m finishing points of E and $(n-m)$ asymptotic directions, $m \le n$.

$$H_{CL}(z)=\frac{0.2kH_{OL}(z)}{1+H_{OL}(z)}$$

$$=\frac{0.2k\dfrac{0.26z+0.3}{(z-1)(z-0.25)}}{1+0.2k\dfrac{0.26z+0.3}{(z-1)(z-0.25)}}$$

$$H_{CL}(z)=\frac{0.26*0.2kz+0.3*0.2k}{(z-1)(z-0.25)+0.26*0.2kz+0.3*0.2k}=\frac{N_{CL}(z)}{D_{CL}(z)}$$

The characteristic equation is

$$D_{CL}(z)=(z-1)(z-0.25)+(0.26z+0.3)0.2k$$
$$=z^2+(0.26*0.2k-1.25)z+0.25+0.3*0.2k$$
$$=z^2+(0.052k-1.25)z+0.25+0.6k$$

Let us consider

$$z=x+jy$$

$$\Rightarrow D_{CL}(z)=(x+jy)^2+(0.052k-1.25)(x+jy)+0.25+0.6k$$
$$=x^2-y^2+(0.052k-1.25)x+0.25+0.6k+jy\left[2x+(0.052k-1.25)\right]$$

$$\begin{cases} k=24-38.46x \\ (x+11.539)^2+y^2=3.4 \end{cases}$$

--2-- MatLab function method:

To find the continuous-time transfer function, we use the MatLab function "tf"

```
>> Hs=tf([1],[1 2 0])
Hs =
     1
  ---------
  s^2 + 2 s

Continuous-time transfer function.
```

The discrete-time transfer function is found by using the MatLab function "c2d"

```
>> Hz=c2d(Hs,1)
Hz =
   0.2838 z + 0.1485
  ---------------------
  z^2 - 1.135 z + 0.1353
 Sample time: 1 second.
Discrete-time transfer function.
```

```
>> Nz=[0.2838 0.1485];
>> Dz=[1 -1.135 0.1353];
```

The discrete-time transfer function using the numerator and denominator:

```
>> Hz=tf(Nz,Dz,-1)
Hz =
   0.2838 z + 0.1485
  ---------------------
  z^2 - 1.135 z + 0.1353
Sample time: unspecified
Discrete-time transfer function.
```

```
>> zpkHz=zpk(Hz)
zpkHz =
  0.2838 (z+0.5233)
  -----------------
  (z-1) (z-0.1353)
 Sample time: unspecified
Discrete-time zero/pole/gain model.
```

The MatLab function that we use to draw the Evans's roots is

```
>> rlocus(Hz),zgrid
```

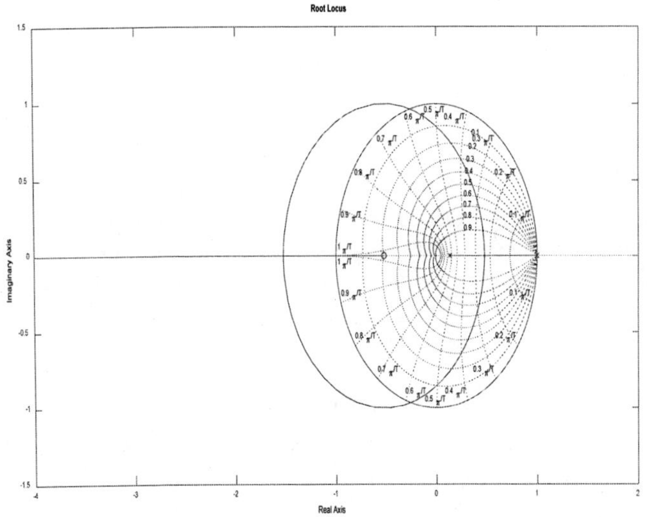

Fig. 4.26. Case of the root locus of the system by using the MatLab function "rlocus"

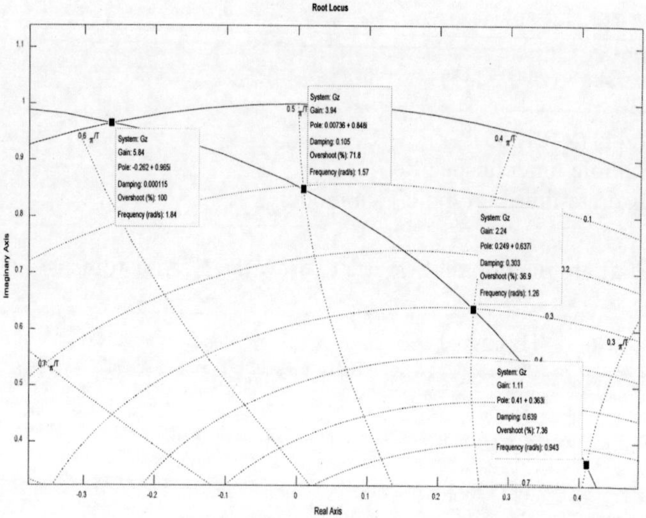

Fig. 4.27. Case of the root locus of the system by using the MatLab function "rlocus"

The performance of a closed loop is noticed by zooming the previous figure. Indeed, when the gain K=1 we verify that damping is equal to 0.6 which corresponds to the overshoot of 4.7% with a frequency that is equal to 0.9rad/s. In the same way, when K=2.24, the damping is 0.3, the overshoot is 37% and frequency is 1.26 rad/s.

The code used to get the selected point in the graphic window is

```
>> [k,poles]=rlocfind(Hz)
Select a point in the graphics window
selected_point =
  0.1388 + 0.7482i
k =
  3.0095
poles =
  0.1404 + 0.7500i
  0.1404 - 0.7500i
```

This is the code used for the comparison study between continuous and discrete closed loop systems.

```
>> clz=feedback(Hz,1)
clz =
    0.2838 z + 0.1485
  ------------------------
  z^2 - 0.8512 z + 0.2838
Sample time: unspecified
Discrete-time transfer function.
>> cls=feedback(Hs,1)
cls =
       1
  -------------
  s^2 + 2 s + 1
Continuous-time transfer function.
```

To draw the unit step response of the continuous and discrete closed loop system, we use this function

```
>> step(clz,cls)
```

The result of the unit step response of the continuous-time and discrete-time closed loop systems are shown in the following figure:

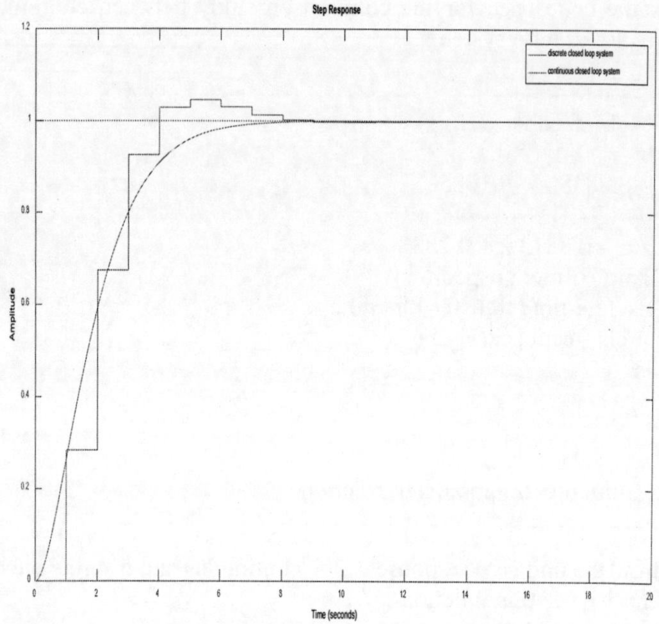

Fig. 4.28. The unit step response of the continuous-time and discrete-time closed loop systems

4.6. Steady state error of discrete-time system

4.6.1. Definition of steady state error

In the closed loop system, as given in the following figure, the steady state error is defined as the difference between the system output, y, and the desired reference signal, r, in the limit as time goes to infinity.

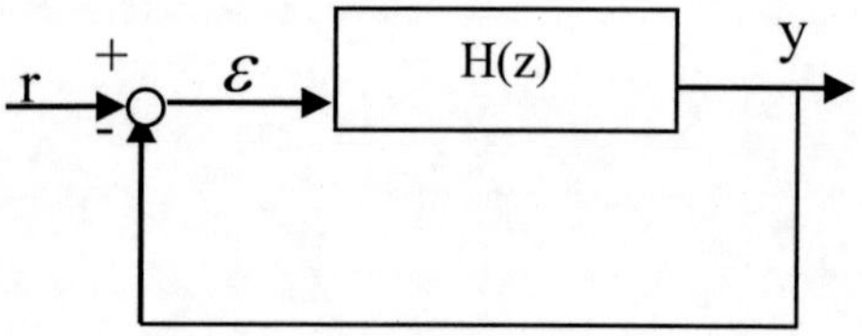

Fig. 4.29. A standard unit feedback discrete-time system

r is the desired performance or reference, ε is the difference error and y is the information about the system or the output.

The expression of the steady state error is

$$e(+\infty) = \lim_{t\to+\infty} \varepsilon(t) = \lim_{z\to 1}(z-1)\varepsilon(z)$$

However,

$$\varepsilon(z) = R(z) - Y(z)$$

and

$$Y(z) = H(z).\varepsilon(z)$$

So we have

$$\varepsilon(z)(1 + H(z)) = R(z)$$

Thus $\varepsilon(z)$ becomes

$$\varepsilon(z) = \frac{R(z)}{1 + H(z)}$$

and $e(+\infty)$ becomes

$$e(+\infty) = \lim_{t\to+\infty} \varepsilon(t) = \lim_{z\to 1}(z-1)\frac{R(z)}{1 + H(z)}$$

yet we have

$$H(z) = \frac{N(z)}{D(z)} = \frac{N(z)}{(z-1)^{\alpha} D_1(z)}$$

so $e(+\infty)$ becomes

$$e(\infty) = \lim_{z\to 1}(z-1)\frac{R(z)}{1 + H(z)}$$

$$= \lim_{z\to 1}(z-1)\frac{R(z)}{1 + \dfrac{N(z)}{(z-1)^{\alpha} D_1(z)}}$$

$$= \lim_{z\to 1}\frac{(z-1)^{\alpha+1} D_1(z)R(z)}{(z-1)^{\alpha} D_1(z) + N(z)}$$

According to the transfer function of $R(z)$, different cases of $e(+\infty)$ are used.

4.6.2. The static position error constant

If the reference signal, $r(k)$, is a unit step, and $R(z)=\dfrac{z}{z-1}$, then by the final value theorem of z-transform, the static position error constant ε_p is

$$\varepsilon_p = e(\infty) = \lim_{z\to 1} \frac{(z-1)^{\alpha} D_1(z)\, z}{(z-1)^{\alpha} D_1(z) + N(z)}$$

According to the value of α, the static position error constant, ε_p, takes different values given in the table below.

Table 4.3. Different values of the static position error

α	ε_p
$\alpha = 0$	$\varepsilon_p = \dfrac{1}{1+H(1)}$
$\alpha > 0$	$\varepsilon_p = 0$

4.6.3. The static velocity error constant

If the reference signal $r(k)$ is a unit ramp, and $R(z)=\dfrac{T_s z}{(z-1)^2}$, then, by the final value theorem of z-transform, the static velocity error constant, ε_v, is

$$\varepsilon_v = e(\infty) = \lim_{z\to 1} \frac{(z-1)^{\alpha} D_1(z) T_s z}{(z-1)^{\alpha} D_1(z) + N(z)}$$

According to the value of α, the static velocity error constant ε_v takes different values given in the table below.

Table 4.4. Different values of the static velocity error

α	ε_V
$\alpha = 0$	$\varepsilon_V = \infty$
$\alpha = 1$	$\varepsilon_V = \dfrac{T_s\, D_1(1)}{N(1)}$
$\alpha > 1$	$\varepsilon_V = 0$

4.6.4. The static acceleration error constant

If the reference signal $r(k)$ is a parabolic input, and $R(z) = \dfrac{T_s^2 z(z+1)}{2(z-1)^3}$, then, by the final value theorem of z-transform, the static acceleration error constant ε_a is

$$\varepsilon_a = e(\infty) = \lim_{z \to 1} \frac{(z-1)^{\alpha-2} D_1(z)\, T_s^2 z(z+1)}{2(z-1)^{\alpha} D_1(z) + N(z)}$$

- If $\alpha < 2$ then

$$\varepsilon_a = \infty$$

- If $\alpha = 2$ then

$$\varepsilon_a = \frac{T_s^2\, D_1(1)}{N(1)}$$

- If $\alpha > 2$ then

$$\varepsilon_a = 0$$

According to the value of α, the static acceleration error constant ε_a takes different values given in the table below.

Table 4.5. Different values of the static acceleration error

R(t) / α	$u(t)$	$t\,u(t)$	$\frac{t^2}{2}u(t)$	$\frac{t^3}{3}u(t)$
0	$\frac{1}{1+H(1)}$	$+\infty$	$+\infty$	$+\infty$
1	**0**	$T_s\frac{1}{1+H(1)}$	$+\infty$	$+\infty$
2	**0**	**0**	$T_s^2\frac{1}{1+H(1)}$	$+\infty$
3	**0**	**0**	**0**	$T_s^3\frac{1}{1+H(1)}$

Example

For a given continuous-time system function $H(s)=\dfrac{K}{s(s+1)}$, using a sampling time $T_s=1$, evaluate, for the discrete-time system, the static error constants and find the expected steady state errors for the standard step, ramp and parabolic inputs (ε_p, ε_v and ε_a).

Answer

The transfer function of the open loop $H_{OL}(z)$ is

$$H_{OL}(z)=\left(1-z^{-1}\right)Z\left[\frac{H(s)}{s}\right]$$

$$=K\frac{z^{-1}\left(1-e^{-T_s}\right)}{1-e^{-T_s}z^{-1}}$$

$$=K\frac{0.37z+0.26}{(z-1)(z-0.37)}$$

Using Tables 4.3, 4.4 and 4.5 we get:

$$\alpha = 1$$

$$\varepsilon_p = 0$$

$$\varepsilon_V = T\frac{D_1(1)}{N_1(1)} = \frac{0.63}{0.63.K} = \frac{1}{K}$$

$$\varepsilon_a = \infty$$

4.7. Applications

Application 1

Consider a 1st order continuous-time system $H(s) = \frac{1}{1+\tau s}$ with a proportional controller K as given as in the following figure.

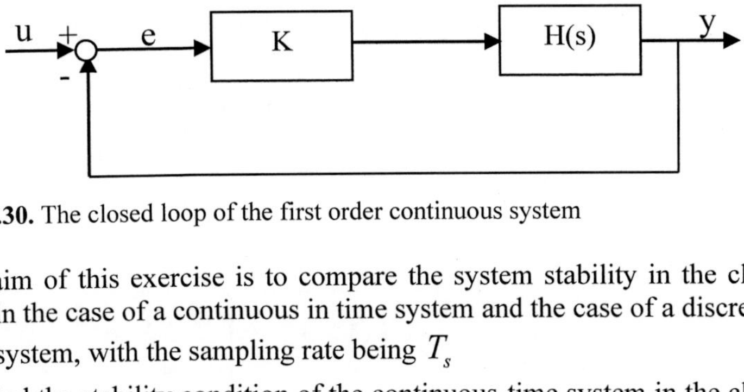

Fig. 4.30. The closed loop of the first order continuous system

The aim of this exercise is to compare the system stability in the closed loop in the case of a continuous in time system and the case of a discrete in time system, with the sampling rate being T_s

1 - Find the stability condition of the continuous-time system in the closed loop.
2 - Find the stability condition of the discrete-time system in the closed loop.
3 - Comment.
4 - Consider $\tau = 10$ and $T_s = 1s$, compute and draw the output response when the input is a unit step if $K = 3$. Find the steady state position error ε_p.
5 - We want that the steady state position error to be zero, what kind of controller could make this possible?

Answer

1- The continuous-time transfer function, in the open loop, is

$$H_{OL}(s)=\frac{Y(s)}{E(s)}=K\frac{1}{1+\tau s}$$

In the closed loop, it is given by:

$$H_{CL}(s)=\frac{Y(s)}{U(s)}=\frac{\frac{1}{1+\tau s}.K}{1+\frac{1}{1+\tau s}.K}=\frac{K}{K+1+\tau s}=\frac{N_{CL}(s)}{D_{CL}(s)}$$

The characteristic equation is

$$D_{CL}(s)=K+1+\tau s=0 \Rightarrow s=-\frac{K+1}{\tau}$$

The unique condition of continuous-time system stability is

$$-\frac{K+1}{\tau}<0 \Rightarrow K>-1$$

2 - The discrete-time transfer function, in the open loop, is

$$H_{OL}(z)=KZ\left[\frac{H(s)}{s}\right]=(1-z^{-1})Z\left[\frac{H(s)}{s}\right]$$

In the closed loop case, it is given by

$$H_{CL}(z)=\frac{Y(z)}{U(z)}=\frac{(1-z^{-1})Z\left[\frac{H(s)}{s}\right].K}{1+(1-z^{-1})Z\left[\frac{H(s)}{s}\right].K}$$

$$H_{CL}(z)=\frac{\frac{z-1}{z}\frac{z(1-e^{-\frac{T_s}{\tau}})}{(z-1)(z-e^{-\frac{T_s}{\tau}})}.K}{1+\frac{z-1}{z}\frac{z(1-e^{-\frac{T_s}{\tau}})}{(z-1)(z-e^{-\frac{T_s}{\tau}})}.K}$$

$$H_{CL}(z)=\frac{(1-e^{-\frac{T_s}{\tau}}).K}{z-e^{-\frac{T_s}{\tau}}+(1-e^{-\frac{T_s}{\tau}}).K}=\frac{N_{CL}(z)}{D_{CL}(z)}$$

The characteristic equation is:

$$D_{CL}(z)=z-e^{-\frac{T_s}{\tau}}+(1-e^{-\frac{T_s}{\tau}}).K \Rightarrow z=e^{-\frac{T_s}{\tau}}-(1-e^{-\frac{T_s}{\tau}}).K$$

The system condition is

$$|z|<1 \Rightarrow -1<e^{-\frac{T_s}{\tau}}-(1-e^{-\frac{T_s}{\tau}}).K<1$$

$$\Rightarrow -1<K<\frac{1+e^{-\frac{T_s}{\tau}}}{1-e^{-\frac{T_s}{\tau}}}$$

3 - According to the given stability condition, we have

$$-1<K<\frac{1+e^{-\frac{T_s}{\tau}}}{1-e^{-\frac{T_s}{\tau}}}$$

We remind that we are looking for the relation between the sampling period T_s and the stability condition, then

$$\Rightarrow K-(1+K)e^{-\frac{T_s}{\tau}}<1 \qquad \Rightarrow -(1+K)e^{-\frac{T_s}{\tau}}<1-K$$

$$\Rightarrow e^{-\frac{T_s}{\tau}}>\frac{1-K}{1+K} \quad \Rightarrow -\frac{T_s}{\tau}>\ln\frac{1-K}{1+K} \quad \Rightarrow T_s<\tau\ln\frac{1-K}{1+K}$$

According to the expression of sampling period $T_s<\tau\ln\frac{1-K}{1+K}$, it is clear that the sampling period depends on the system parameters.

4 - Consider $\tau=10$ and $T_s=1s$, and we compute the response step of the system if $K=3$:

$$H_{CL}(z)=\frac{(1-e^{-\frac{T_s}{\tau}}).K}{z-e^{-\frac{T_s}{\tau}}+(1-e^{-\frac{T_s}{\tau}}).K}=\frac{Y(z)}{U(z)}$$

$$\Rightarrow Y(z)=\frac{(1-e^{-\frac{T_s}{\tau}}).K}{z-e^{-\frac{T_s}{\tau}}+(1-e^{-\frac{T_s}{\tau}}).K}U(z)$$

$$Y(z)=\frac{0.2z}{(z-1)(z-0.6)}$$

We find simple elements of:

$$\frac{Y(z)}{z}=\frac{a}{z-1}+\frac{b}{z-0.6}$$

With

$$a=\lim_{a\to 1}(z-1)\frac{Y(z)}{z}=\lim_{a\to 1}\frac{0.2}{z-0.6}=0.5$$

$$b=\lim_{b\to 0.6}(z-0.6)\frac{Y(z)}{z}=\lim_{a\to 0.6}\frac{0.2}{z-1}=-0.5$$

we get

$$Y(z)=\frac{0.5z}{z-1}-\frac{0.5}{z-0.6}\Rightarrow y(k)=0.5(1^k-0.6^k)u(k)$$

k	0	1	2	3	4	5	7	9	10
$y(k)$	0	0.2	0.32	0.39	0.43	0.46	0.48	0.49	0.49

Finally, the steady state position error is

$$\begin{aligned}\varepsilon_p &= \lim_{z\to 1}(z-1)E(z)\\ &= \lim_{k\to\infty}\left[u(k)-y(k)\right]\\ &= \lim_{k\to\infty}\left[u(k)-0.5(1^k-0.6^k)u(k)\right]\\ &=1-0.5=50\%\end{aligned}$$

5 - To find a steady state position error as zero, we should use an integral controller.

Application 2

Let us consider the following figure.

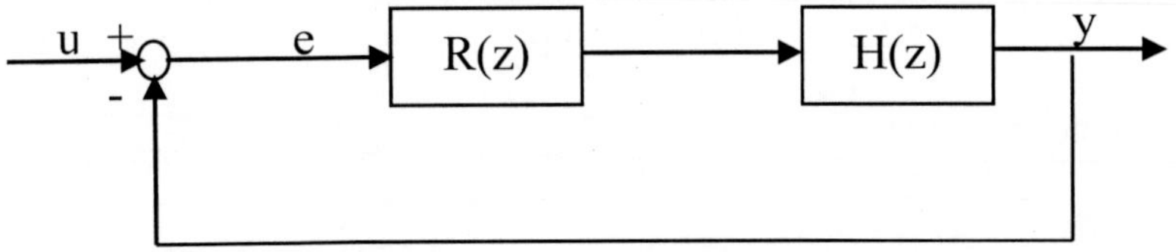

Fig. 4.31. The closed loop of the discrete-time system

The discrete transfer function is $H(z) = \dfrac{z+0.1}{(z-1)(z-0.4)}$ and R(z) is a proportional controller with gain K.

1 - Find the condition stability using the Routh criterion.
2 - Find the condition stability using the Jury criterion.
3 - Find the steady state position error.
4 - Find the steady state velocity error. How can we decrease this error without changing the controller. What kind of problems could occur?

Answer

1 - In the closed loop, the transfer function is

$$H_{CL}(z) = \frac{K..H(z)}{1+K.H(z)}$$

$$= \frac{K(z+0.1)}{(z-1)(z-0.4)+K(z+0.1)}$$

$$H_{CL}(z) = \frac{K(z+0.1)}{z^2+(K-1.4)z+0.4+0.1K} = \frac{N_{CL}(z)}{D_{CL}(z)}$$

The characteristic equation is

$$D_{CL}(z) = z^2+(K-1.4)z+0.4+0.1K$$

By applying the Routh criterion, we get

$$D_{CL}(z) = z^2+(K-1.4)z+0.4+0.1K$$

$$D_{CL}(z = \frac{1+w}{1-w}) = (\frac{1+w}{1-w})^2+(K-1.4)(\frac{1+w}{1-w})+0.4+0.1K$$

$$D_{CL}(\frac{1+w}{1-w})=(\frac{1+w}{1-w})^2+(K-1.4)\frac{(1+w)(1-w)}{(1-w)^2}+(0.4+0.1K)\frac{(1-w)^2}{(1-w)^2}$$

$$(1-w)^2 D_{CL}(\frac{1+w}{1-w})=w^2(2.8-0.9K)+w(1.2-0.2K)+1.1K$$

By the following table of the Routh criterion, we see

w^2	2.8-0.9K	1.1K
w^1	1.2-0.2K	0
w^0	1.1K	0

We will now check the stability conditions:

$$\left.\begin{array}{ll} 1.1K>0 & \Rightarrow K>0 \\ (2.8-0.9K)>0 & \Rightarrow K<3.11 \\ (1.2-0.2K)>0 & \Rightarrow K<5.5 \end{array}\right\} \Rightarrow \quad 0<K<3.11$$

2 - In this case, we want to apply the Jury criterion using the characteristic equation

$$D(z)=z^2+(K-1.4)z+0.4+0.1K$$

The system is stable if

$$D(1)=1+K-1.4+0.4+0.1K=1.1K>0 \quad \Rightarrow \quad K>0$$

$$D(-1)=1-K+1.4+0.4+0.1K=2.2-0.9K>0 \quad \Rightarrow \quad K<2.44$$

$$|0.4+0.1K|<1 \Rightarrow -1<0.4+0.1K<1 \Rightarrow -14<K<60$$

3 - The closed loop transfer function is

$$H_{CL}(z)=\frac{K(z+0.1)}{z^2+(K-1.4)z+0.4+0.1K}=\frac{Y(z)}{U(z)}$$

$$\Rightarrow Y(z)=H_{CL}(z)U(z)$$

$$=\frac{K(z+0.1)}{z^2+(K-1.4)z+0.4+0.1K}\frac{z}{z-1}$$

The steady state position error is

$$\varepsilon_p = \lim_{z\to 1}(z-1)\frac{U(z)}{1+K.H(z)}$$

$$= \lim_{z\to 1}(z-1)\frac{\dfrac{z}{z-1}}{\dfrac{(z-1)(z-0.4)}{(z-1)(z-0.4)} + K\dfrac{z+0.1}{(z-1)(z-0.4)}}$$

$$= \lim_{z\to 1}(z-1)\frac{z(z-0.4)}{(z-1)(z-0.4)+K(z+0.1)} = 0$$

4 - The steady state velocity error is

$$\varepsilon_v = \lim_{z\to 1}(z-1)\frac{U(z)}{1+K.H(z)}$$

$$= \lim_{z\to 1}(z-1)\frac{\dfrac{z}{(z-1)^2}}{\dfrac{(z-1)(z-0.4)}{(z-1)(z-0.4)} + K\dfrac{z+0.1}{(z-1)(z-0.4)}}$$

$$= \lim_{z\to 1}\frac{z(z-0.4)}{(z-1)(z-0.4)+K(z+0.1)}$$

$$= \frac{0.54}{K}$$

We notice that we can increase the gain of the proportional controller K to decrease the steady state velocity error. We can improve the steady state velocity error ($\varepsilon_v \to 0$), but K can be higher than 3.11 and the system becomes instable.

4.8 Conclusions

In this chapter we presented the stability condition of discrete-time systems. Indeed, the algebraic stability criterion and the graphical stability criterion are detailed. The advantage and the disadvantage of each criterion is shown. However, the last section of this chapter treated the system performance.

4.9 References

Agarwal, R.C. and Burrus, C.S. 1975. New recursive digital filter structures having very low sensitivity and roundoff noise. IEEE Transactions on Circuits and Systems, Vol. 22, N° 12: 921–927.

Bruce, P. 2017. Essential process control for chemical engineers: bookboon.com ISBN 978-87-403-1655-1.

Arthur, G.O.M. 1999. Design and analysis of control systems, CRC Press.

Astrom K.J. and Wittenmark B. 1997. Computer-controlled systems: theory and design, Prentice Hall.

Borne, P., Dauphin-Tanguy, G., Richard, J.P., Rotella, F. and Zambettakis, I. 1993. Analyse et régulation des processus industriels. Tome 1: régulation continue. Technip, France.

Borne, P., Dauphin-Tanguy, G., Richard, J.P., Rotella, F. and Zambettakis, I. 1993. Analyse et régulation des processus industriels. Tome 2: régulation numérique. Technip, France.

Brogan, W. 1991. Modern control theory. Prentice Hall.

Arild, W. 2013. Discrete dynamical systems with an introduction to discrete optimization problems: bookboon.com ISBN 978-87-403-0327-8.

Bishop, R. 2002. The control handbook. The electrical engineering handbook series. CRC Press/IEEE Press.

Cheng, H.-M. and Chiu, G.T.-C. 2007. Coupling between sample rate and required word length for finite precision controller implementation with delta transform: 3588–3593.

Dorf, R.C. and Bishop, R.H. 1995. Modern control systems. Addison-Wesley Publishing Company, Inc., New-York.

Shankar, S. and Marc, B. 1989. Adaptive control: stability, convergence, and robustness: Prentice Hall information and system sciences series.

Dorf, R.C. and Bishop, R.H. 2008. Modern control systems, Pearson Prentice Hall.

Dieulesaint, E. and Royer, D. 1987. Automatique appliquée: 1. systèmes linéaires de commande à signaux analogiques. Masson, France.

Dieulesaint, E. and Royer, D. 1990. Automatique appliquée: 2. systèmes linéaires de commande à signaux échantillonnés. Masson, France.

Feuer, A. and Goodwin, G.C. 1996. Sampling in digital signal processing and control. Birkhauser.

Franklin, G.F., Powell, J.D. and Emami-Naeini, A. 1994. Feedback control of dynamic systems. Addison-Wesley Publishing Company, Inc., New-York.

Derek, A. 2014. An introduction to nonlinearity in control systems: bookboon.com ISBN 978-87-7681-790-9.
Franklin, G.F., Powell, J.D. and Workman, M.L. 1997. Digital control of dynamic systems. Addison Wesley, 3rd edition.
Franklin, G.F., Powell, J.D. and Workman, M.L. 1998. Digital control of dynamic systems, 3rd Edition, Addison Wesley.
Gajić, Z. 2003. Linear dynamic systems and signals. Prentice Hall.
Goodwin, G.C. 2006. Efficient data representations for signal processing and control: making most of a little. Proceedings of the 25th Chinese Control Conference: 19- 37.
Goodwin, G.C., Graebe, S.F. and Salgado, M.E. 2001. Control systems design. Prentice Hall.
Gang, F. and Rogelio, L. 1999. Adaptive control systems. Butterworth-Heinemann Linacre House, Jordan Hill, Oxford OX2 8DP.
Hanslemann, H. 1987. Implementation of digital controllers, a survey. Automatica, Vol. 23, N° 1: 7-32.
Jaume, D., Thelliez, S. and Vergé, M. 1991. Commande des systèmes dynamiques par calculateur. Eyrolles, France.
Lnooka, H., Obinata, G. and Takeshima, M. 1983. Design of digital controller based on series expansion of pulse transfer functions, Journal of Dynamic Systems, Measurements and Control, Vol. 105, N°. 3: 204-207.
Nishimura, S., Hirano, K., and Pal, R.N. 1981. A new class of very low sensitivity and low roundoff noise recursive digital filter structures. IEEE Trans. Circuits Syst., CAS-28(12): 1152-1158.
Derek, A. 2015. Control engineering problems with solutions: bookboon.com ISBN 978-87-403-0374-2.
Pradin, B. 1999. Systèmes à temps discret- commande numérique des procédés. INSA Toulouse, France.
Oppenheim, A.V. and Schafer, R.W. 1998. Discrete-time signal processing. Prentice-Hall, 2 edition.
Oppenheim, A.V., Willsky, A.S. and Hamid, S. 1996. Signals and systems, 2nd edition, Prentice-Hall.
Richard, H.M. and Goodwin, G.C. 1986. Improved finite word length characteristics in digital control using delta operators. IEEE Trans. on Automatic Control, AC-31: 1015-1021.
Steinbuch, M., Schoostra, G., and Goh, H.T. 1994. Closed-loop scaling in fixed-point digital control. IEEE Trans. on Control Systems Tech., 2: 312-317.
Sami, F.M. and Antonio, V. 2009. Digital control engineering: analysis and design academic press.

Tiwari, J.P. and Bhagat, S.K. 2004. Simplification of discrete time systems by improved Routh stability criterion via P-domain, Journal of IE, Vol. 85: 89-192.

Zhengtao, D. 2013. Nonlinear and adaptive control systems: The Institution of Engineering and Technology.

CHAPTER FIVE

SYNTHESIS OF DIGITAL CONTROLLER FOR LINEAR TIME INVARIANT SYSTEM

5.1. Introduction

In this chapter, we mainly consider the control system, namely, the Proportional-Integral-Derivative (PID) controller. PID controllers are a versatile category of controllers that are commonly used in industry as control systems due to the ease of their implementation and low cost. One problem that continues to intrigue control designers is the matter of finding a good combination of the three parameters K_P, K_I and K_D of these controllers so that system stability and optimum performance is achieved. Also, a certain amount of robustness to the process is expected from the PID controllers.

5.2 Digital controllers

5.2.1 Proportional action

The control law $u_P(t)$, in the continuous-time case, is

$$u_P(t) = K_P \varepsilon(t)$$

and the transfer function of the continuous proportional controller is:

$$C_P(s) = \frac{U_P(s)}{\varepsilon(s)} = K_P$$

where $\varepsilon(t)$ is the error signal as given in the feedback block diagram and K_P is the gain of the proportional controller. In the following figure, the continuous-time control system using a proportional controller is shown.

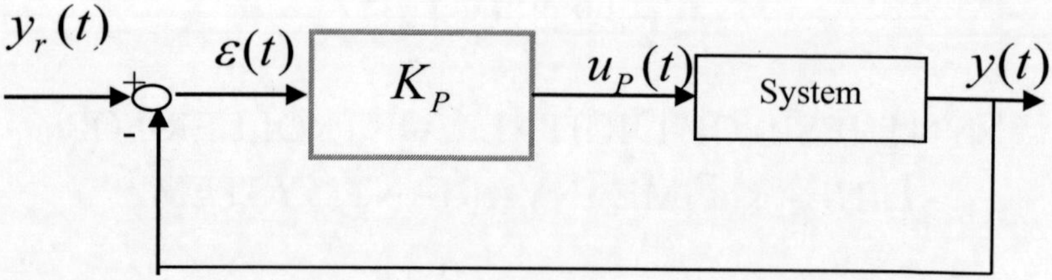

Fig. 5.1. The continuous-time control system using a proportional controller

with $y_r(k)$ is a desired value or a signal reference, y(t) is the system output and $u_p(t)$ is the control law.

In the discrete-time case, the control law is

$$u_P(k) = K_P\varepsilon(k) = K_P\left(y_r(k) - y(k)\right)$$

The transfer function of the discrete proportional controller is:

$$C_P(z) = \frac{U_P(z)}{\varepsilon(z)} = K_P$$

The proportional gain K_P determines how fast the controller will react to the error input ε, indeed, too low a value will make the controller react slowly but too high of a value may make the system unstable.

Example: The following figure shows the response of the output to a unit step in the command signal for a system with a pure proportional controller. The system has the transfer function $H(z)$. Three cases of the proportional controller parameters ($K_P = 1$, 2 and 4) are taken to show their influence in the system output.

The transfer function is given by the following expression.

$$H(z) = \frac{0.4512}{z - 0.5488}$$

The closed loop discrete-time system using a proportional controller is presented in the following figure.

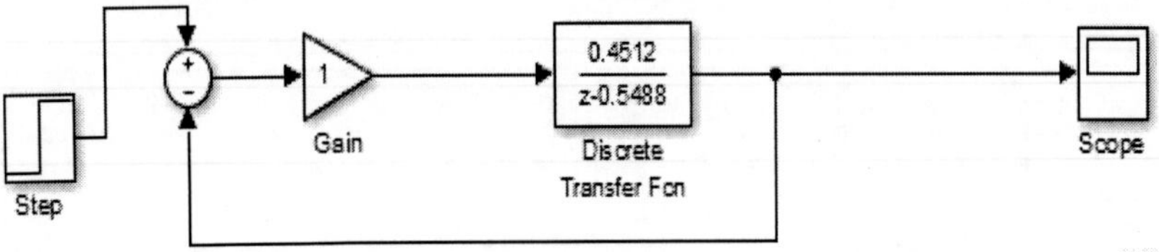

Fig. 5.2. The discrete-time control system using a proportional controller

Using the Simulink/MatLab software, the influence made by the value of the proportional gain of the tracking system output to the desired value is shown by the following figures.

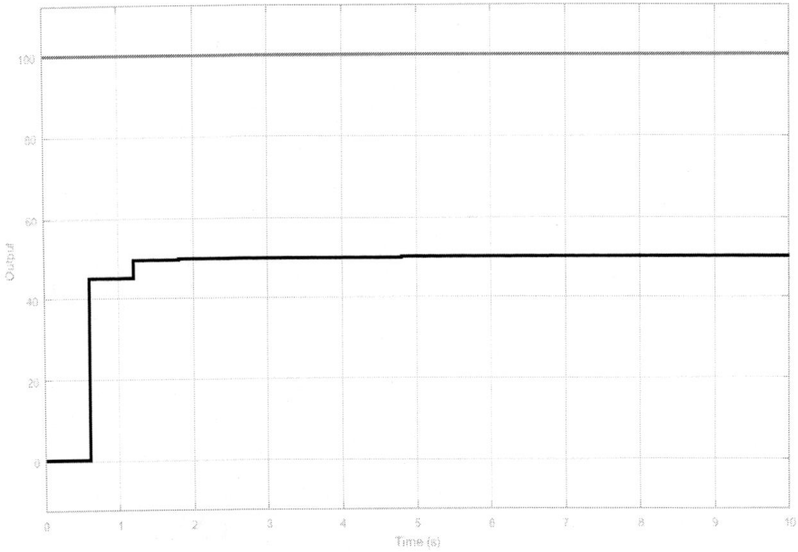

a) The proportional gain $K_P = 1$

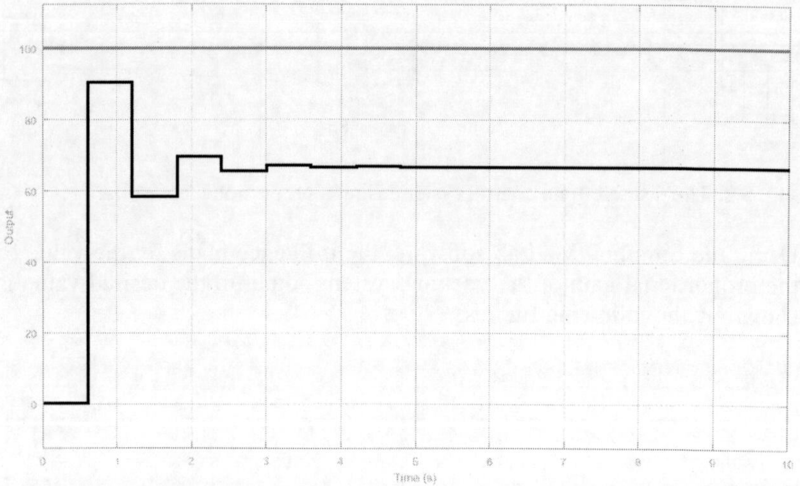

b) The proportional gain $K_P = 2$

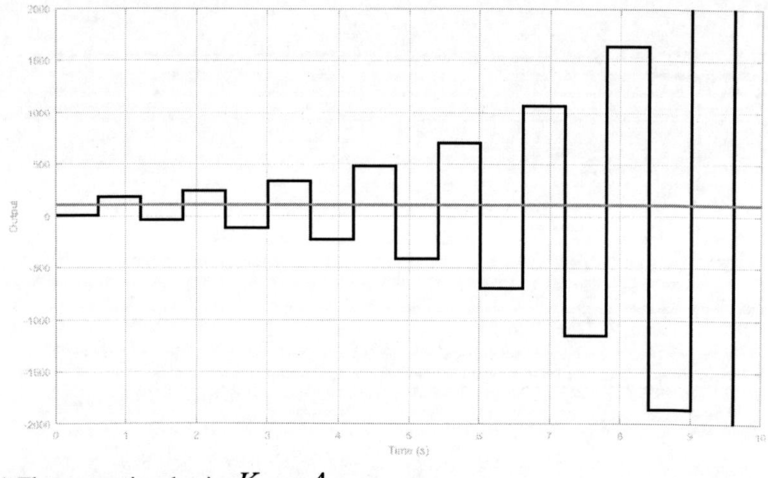

c) The proportional gain $K_P = 4$

Fig. 5.3. Responses (Black) to step (Blue) changes in the command signal for a proportional controller.

According to these figures we remark that the system output never reaches the steady state error. The error decreases with increasing gain ($K_P = 1$, $K_p = 2$ and $K_p = 4$), but the system also becomes more oscillatory. Notice, in the figure, that the initial value of the control signal equals the controller gain.

The main usage of the proportional action controller is to decrease the steady state error of the system. If the proportional gain factor K_P increases, then the steady state error of the system decreases. However, despite the reduction, the proportional action controller can never manage to eliminate the steady state error of the system. As we increase the proportional gain, it provides smaller amplitude and phase margin, faster dynamics satisfying wider frequency band and larger sensitivity to the noise.

In addition, it can be easily concluded that applying a proportional action controller decreases the rise time and after a certain value of reduction on the steady state error, increasing K_P only leads to an overshoot of the system response.

5.2.2 Integral action

Integral action guarantees that the system output agrees with the reference in steady state.

In the continuous case, the control law $u_I(t)$ is

$$u_I(t) = \frac{K_P}{T_I}\int_{t_0}^{t} \varepsilon(t)dt$$

$$= K_I \int_{t_0}^{t} \varepsilon(t)dt$$

and its derivative form is

$$\frac{du_I(t)}{dt} = K_I \varepsilon(t)$$

with $K_I = \dfrac{K_P}{T_I}$ an integral gain and T_I an integral time.

The transfer function of the continuous integral controller is:

$$C_I(s) = \frac{U_I(s)}{\varepsilon(s)} = \frac{K_I}{s}$$

In the following figure the control system using an integral controller is shown.

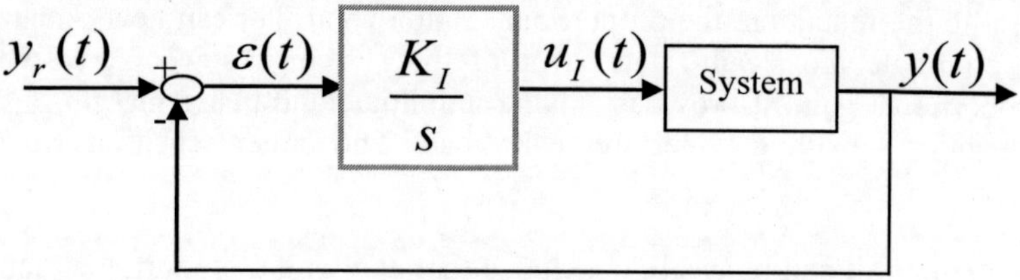

Fig. 5.4. The continuous-time control system using an integral controller

In the following table we show the equivalence between s and z :

Table 5.1. Different method of z equivalence

Method	s to z equivalence
Euler's forward method	$s = \dfrac{1-z^{-1}}{T_s z^{-1}}$
Euler's backward method	$s = \dfrac{1-z^{-1}}{T_s}$
Tustin method	$s = \dfrac{2}{T_s}\dfrac{1-z^{-1}}{1+z^{-1}}$

According to this table and using Euler's method, $\dfrac{d}{dt}$ is similar to $\dfrac{1-z^{-1}}{T_s}$ and $\int dt$ is similar to $\dfrac{T_s}{1-z^{-1}}$, thus the discrete control law is:

$$\frac{u_I(kT_s) - z^{-1}u_I(kT_s)}{T_s} = \frac{1 - z^{-1}}{T_s} u_I(kT_s) = K_I \varepsilon(kT_s)$$

The expression of the control law $u_I(kT_s)$ is:

$$u_I(kT_s) = K_I \frac{T_s}{1 - z^{-1}} \varepsilon(kT_s) = K_I \frac{zT_s}{z - 1} \varepsilon(kT_s)$$

and the transfer function of the integral controller is

$$C_I(z) = \frac{U_I(z)}{\varepsilon(z)} = K_I \frac{T_s z}{z - 1} = K_I \frac{T_s}{1 - z^{-1}}$$

Then, using the expression of the transfer function, the control law is

$$u_I(k) = u_I(k-1) + K_I T_s \left(y_r(k) - y(k)\right)$$

The integral gain K_I compensates for accumulated errors and thus determines overshoot.

Example: The following figure shows the response of the output to a unit step in the command signal for a system with a pure integral action controller. The actual case is the previous transfer function of the system. The integral action controller parameters are $T_s = 0.6s$, $K_I = 0.1$, $K_I = 0.5$ and $K_I = 2$ for the integral action controller.

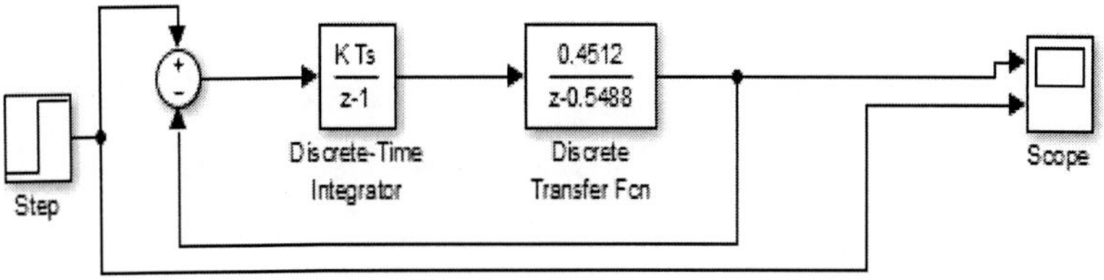

Fig. 5.5. The discrete-time control system using an integral action controller

As given in the case of proportional action, the influence made by the value of the integral gain of the tracking system output to the desired value is shown by the following figures.

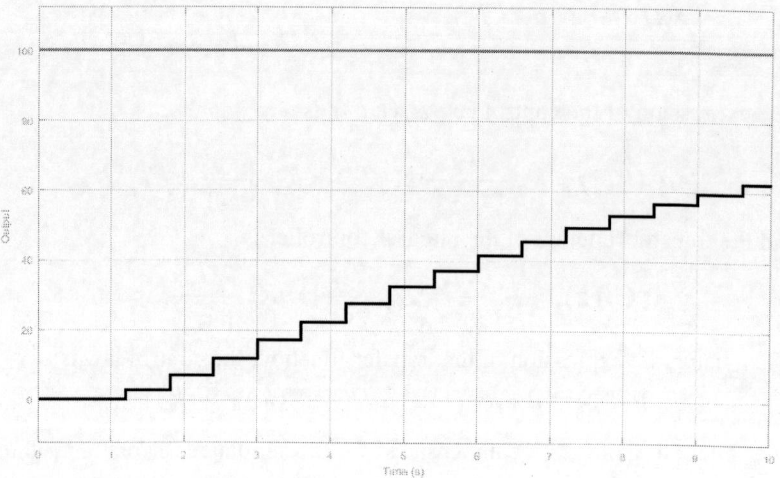
Output
Time (s)
0
20
40
60
80
100
0
1
2
3
4
5
6
7
8
9
10

a) $K_I = 0.1$

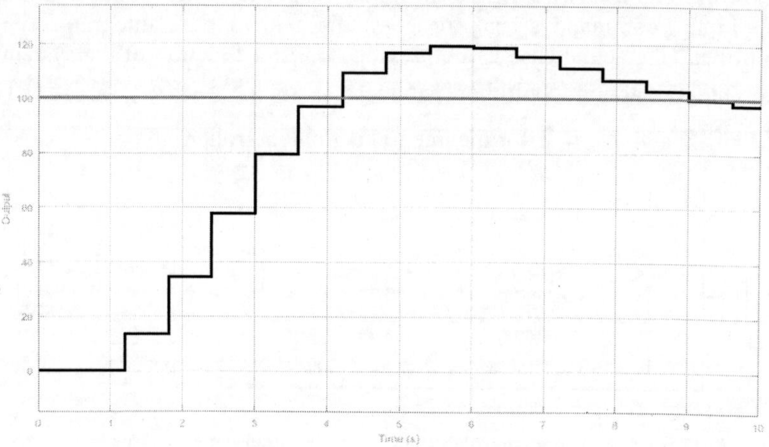
Output
Time (s)
0
20
40
60
80
100
120
0
1
2
3
4
5
6
7
8
9
10

b) $K_I = 0.5$

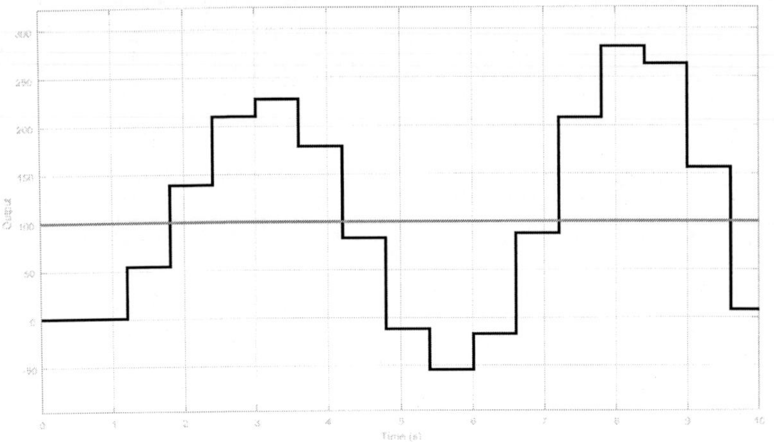

c) $K_I = 2$

Fig. 5.6. Responses (Black) to step (Blue) changes in the command signal for integral action controller.

The steady state error is removed when integral gain K_I is increased. The response creeps slowly towards the reference for small values of K_I. The approach is faster for larger integral gains but the system also becomes more oscillatory.

5.2.3 Derivative action

The derivative action can improve the stability of the closed-loop system. The derivative control is

$$u_D(t) = K_P T_D \frac{d\varepsilon(t)}{dt} = K_D \frac{d\varepsilon(t)}{dt}$$

with $K_D = K_P T_D$ is a derivative gain, T_D is a derivative time and the coefficient of derivative and the transfer function of the continuous derivative action controller is:

$$C_D(s) = \frac{U_D(s)}{\varepsilon(s)} = K_D \frac{d\varepsilon(t)}{dt}$$

In the following figure the control system using a derivative action controller is shown.

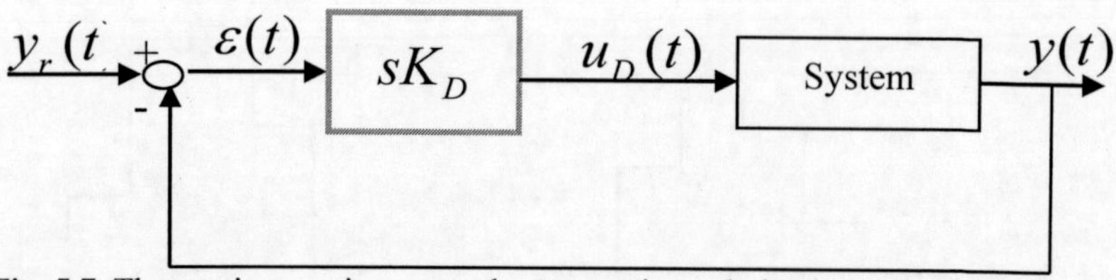

Fig. 5.7. The continuous-time control system using a derivative action controller

Using Euler's method, the transfer function is

$$C_D(z) = \frac{U(z)}{\varepsilon(z)} = K_D \frac{1 - z^{-1}}{T_s} = K_D \frac{z-1}{T_s z}$$

the control law is

$$u_D(k) = \frac{K_D}{T_s}\big(y_r(k) - y(k)\big) - \frac{K_D}{T_s}\big(y_r(k-1) - y(k-1)\big)$$

The derivative gain K_D slows the overshoot but is very sensitive to noise and can cause the system to become unstable due to it.

Derivative action is usually used to improve transient response of the closed loop system. Only derivative action is not used because it amplifies high frequency noise which is never desired. Derivative action decreases rise time and oscillations. However, it does not have any effect on steady state performance of the closed loop.

5.2.4 Filtering the derivative action

A drawback with derivative action is that an ideal derivative has very high gain for high frequency signals. This means that high frequency measurement noise will generate large variations of the control signal.

The effect of measurement noise be reduced by replacing the term $\mathrm{sK_D}$ in the transfer function by

$$\mathrm{C_{DF}(s) = \frac{U_D(s)}{\varepsilon(s)} = K_D \frac{s}{1 + \frac{T_D}{N} s}}$$

and the control law is:

$$U_{DF}(s) = K_D \frac{s}{1+\frac{T_D}{N}s}\varepsilon(s)$$

with $\left(1+\frac{T_D}{N}s\right)$ is the effect of the filtering.

In the following figure the control system using a derivative action controller with filter is shown.

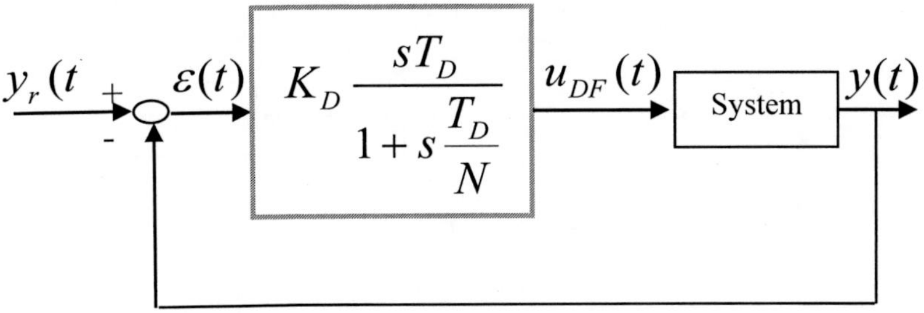

Fig. 5.8. The continuous-time control system using derivative action controller with filter

In the discrete-time case the control law is

$$U_{DF}(z) = K_D \frac{\frac{1-z^{-1}}{T_s}}{1+\frac{T_D}{N}\frac{1-z^{-1}}{T_s}}\varepsilon(z)$$

In the compact form, the control law is

$$U_{DF}(z) = K_D \frac{N(1-z^{-1})}{(N\frac{T_s}{T_D}+1)-z^{-1}}\varepsilon(z)$$

$$= -K_D Np\frac{(1-z^{-1})}{1+pz^{-1}}\varepsilon(z)$$

with $p = \frac{-1}{1+N\frac{T_s}{T_D}}$

5.2.5 Proportional-Integral controller

In this case, the transfer function in the continuous-time is

$$C_{PI}(s) = \frac{U_{PI}(s)}{\varepsilon(s)} = C_P(s) + C_I(s) = K_P + \frac{K_P}{T_I}\frac{1}{s} = K_P + \frac{K_I}{s}$$

In the following figure the control system using a proportional integral controller.

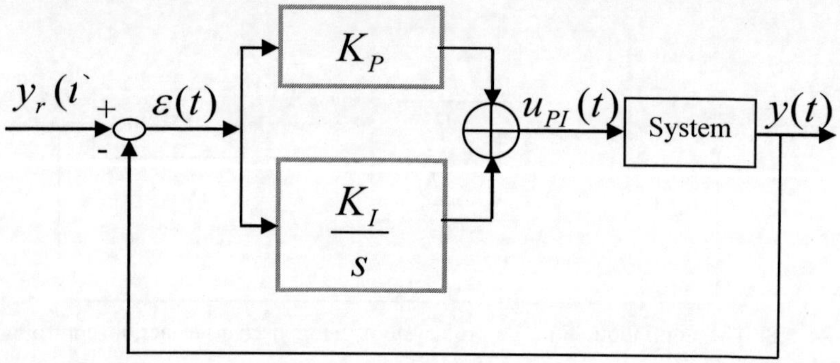

Fig. 5.9. The continuous-time control system using a proportional integral controller

In the discrete-time case, the transfer function is

$$C_{PI}(z) = \frac{U_{PI}(z)}{\varepsilon(z)} = K_P + K_I \frac{T_s}{1 - z^{-1}}$$

This controller improves performances of system control.

Example: The following figure shows the response of the output to a unit step in the command signal for a system with pure proportional integral control using the previous example of transfer function of the system. The controller parameters are $\{K_P = 1;\ K_I = 0.1\}$, $\{K_P = 1;\ K_I = 0.5\}$ and $\{K_P = 1;\ K_I = 1\}$ for the proportional-integral controller.

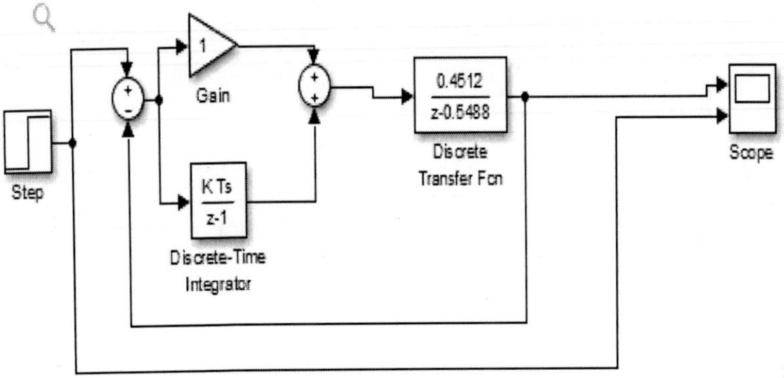

Fig. 5.10. The discrete-time control system using a proportional integral action controller

In this case, the influence made by the value of the proportional gain and the integral gain of the tracking system output to the desired value is shown by the following figures.

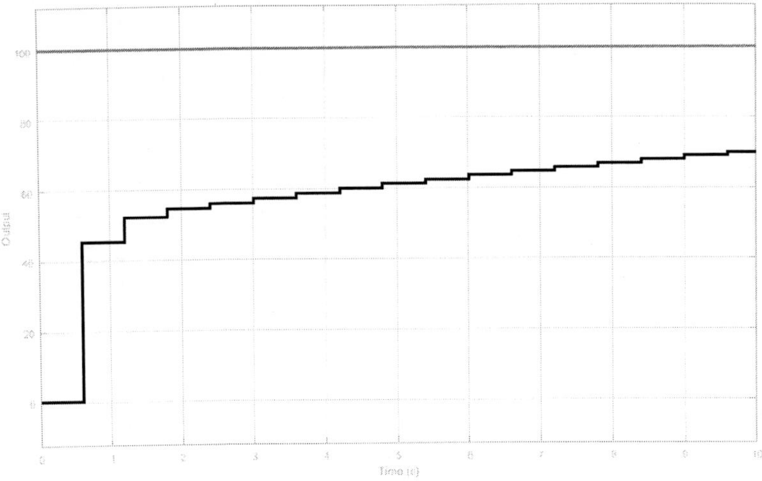

a) $K_P = 1$ and $K_I = 0.1$

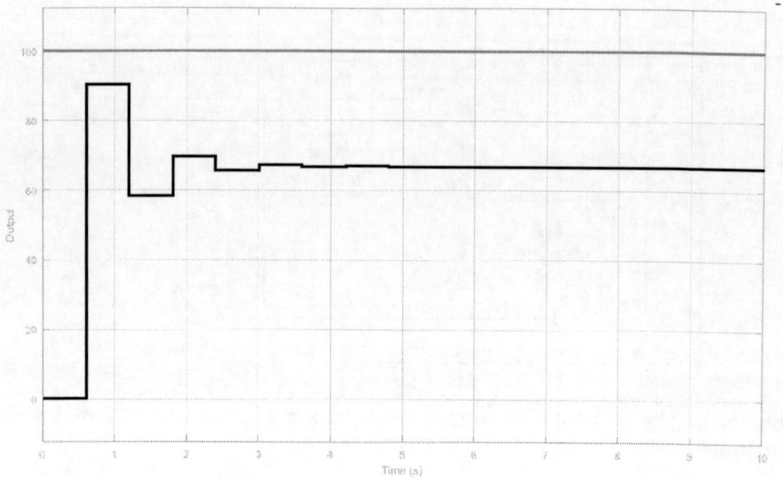

b) $K_P = 1$ and $K_I = 0.5$

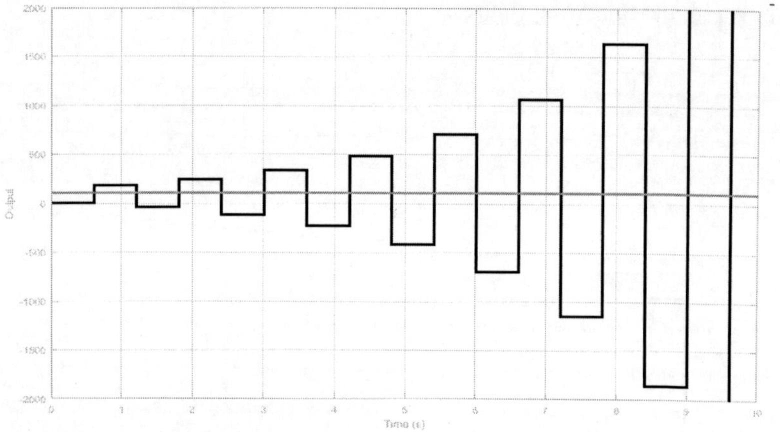

c) $K_P = 1$ and $K_I = 1$

Fig. 5.11. Responses (Black) to step (Blue) changes in the command signal for proportional integral controller.

The proportional integral controller is mainly used to eliminate the steady state error resulting from the proportional action. However, in terms of the speed of the response and overall stability of the system, it has a negative impact.

This controller is mostly used in areas where speed of the system is not an issue. Since the proportional integral controller has no ability to predict the future errors of the system it cannot decrease the rise time and eliminate the oscillations. If applied, any amount of integral action guarantees set point overshoot.

5.2.6 Proportional-Derivative controller

In this case, the transfer function in the continuous-time is

$$C_{PD}(s) = \frac{U_{PD}(s)}{\varepsilon(s)} = C_P(s) + C_D(s) = K_P + sK_P T_D = K_P(1 + sT_D)$$

with $K_D = K_P T_D$.

In the following figure the control system uses a proportional derivative controller.

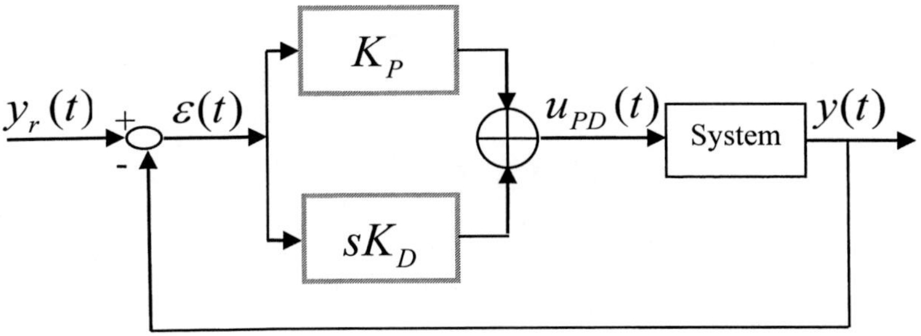

Fig. 5.12. The continuous-time control system using a proportional-derivative controller

In the discrete-time the transfer function is

$$C_{PD}(z) = \frac{U_{PD}(z)}{\varepsilon(z)} = K_P + K_D \frac{1 - z^{-1}}{T_s}$$

This controller improves performances of system control. Indeed, the aim of using proportional derivative action controller is to increase the stability of the system by improving control since it has an ability to predict the future error of the system response. To avoid effects of the sudden change in the value of the error signal, the derivative is taken from the output response of the system variable instead of the error signal. Therefore, derivative action mode is designed to be proportional to the change of the output variable to prevent the sudden changes occurring in the control output resulting from sudden changes in the error signal. In addition, derivative action directly amplifies process noise therefore derivative action only control is not used.

5.3 Proportional-Integral-Derivative controller

The transfer function of the continuous PID controller is

$$C_{PID}(s)=\frac{U_{PID}(s)}{\varepsilon(s)}=C_P(s)+C_D(s)+C_I(s)=K_P\left[1+\frac{1}{T_I s}+T_D s\right]$$

This equation shows the transfer function of the controller in continuous-time. Although, for computational purposes and for practical implementation it is a general practice to put a derivative pseudo-pole of the form $\left(1+\frac{T_D}{N}s\right)$, as a denominator to the derivative term, to ensure that the frequency response rolls-off at high frequency. Thus, the transfer function takes the form shown in the next equation.

$$C_{PID}(s)=\frac{U_{PID}(s)}{\varepsilon(s)}=C_P(s)+C_{DF}(s)+C_I(s)=K_P\left[1+\frac{1}{T_I s}+\frac{T_D s}{1+\frac{T_D}{N}s}\right]$$

The following figure shows the principle of the proportional integral filtering derivative controller.

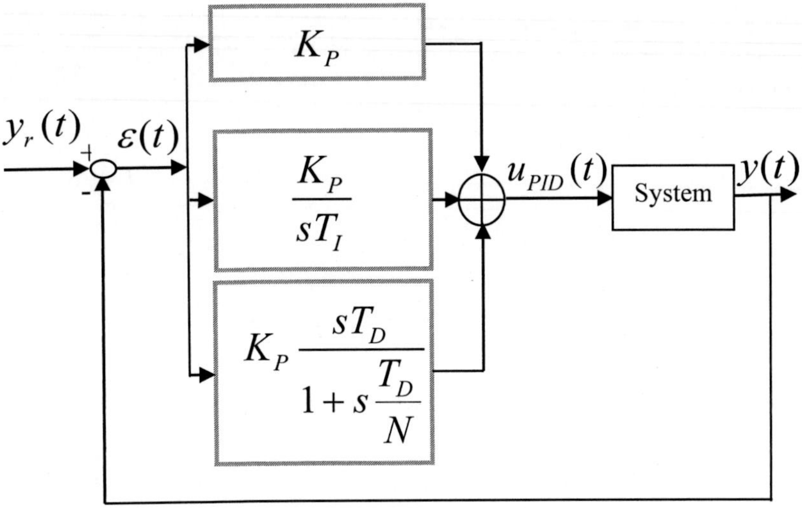

Fig. 5.13. The continuous-time control system using a proportional-integral-filtering-derivative controller

The new control law is

$$U_{PID}(s) = K_P\left[1 + \frac{1}{sT_I} + \frac{sT_D}{1 + s\frac{T_D}{N}}\right]\varepsilon(s)$$

Using Euler's method, the control law becomes

$$U_{PID}(z) = \left[K_P + \frac{K_P T_s}{T_I}\frac{1}{1 - z^{-1}} - \frac{K_P pN(1 - z^{-1})}{1 + pz^{-1}}\right]\varepsilon(z)$$

with $p = -\frac{1}{1 + \frac{NT_s}{T_D}}$

In order to let the expression of the control law clear let's make one denominator as given

$$U_{PID}(z)=K_P\frac{\left(1-z^{-1}\right)\left(1+pz^{-1}\right)+\frac{T_s}{T_I}\left(1+pz^{-1}\right)-pN\left(1-z^{-1}\right)}{\left(1-z^{-1}\right)\left(1+pz^{-1}\right)}\varepsilon(z)$$

and in the next form

$$U_{PID}(z)=K_P\frac{\left(1-Np+\frac{T_s}{T_I}\right)-z^{-1}\left(1-p\left(1+N+\frac{T_s}{T_I}\right)\right)-z^{-2}p}{\left(1-z^{-1}\right)\left(1+pz^{-1}\right)}\varepsilon(z)$$

Finally, the expression of the control law becomes

$$U_{PID}(z)=\frac{r_0+r_1z^{-1}+r_2z^{-2}}{\left(1-z^{-1}\right)\left(1+pz^{-1}\right)}\varepsilon(z)=\frac{R(z)}{S(z)}\varepsilon(z)$$

with

$$R(z)=r_0+r_1z^{-1}+r_2z^{-2}$$
$$S(z)=\left(1-z^{-1}\right)\left(1+pz^{-1}\right)$$
$$\varepsilon(z)=Y_r(z)-R(z)$$

and

$$\begin{cases} r_0=K_P(1-pN+\frac{T_s}{T_I}) \\ r_1=-K_P\left[1-p(1+2N+\frac{T_s}{T_I})\right] \\ r_2=-pK_P \\ p=-\dfrac{1}{1+\dfrac{NT_s}{T_D}} \end{cases}$$

By using $R(z)$ and $S(z)$ in the expression of the control law $U_{PID}(z)$, then it becomes

$$U_{PID}(z) = \frac{R(z)}{S(z)} Y_r(z) - \frac{R(z)}{S(z)} Y(z)$$

In the following figure, the control system using a proportional integral filtering-derivative controller using $R(z^{-1})$ and $S(z^{-1})$.

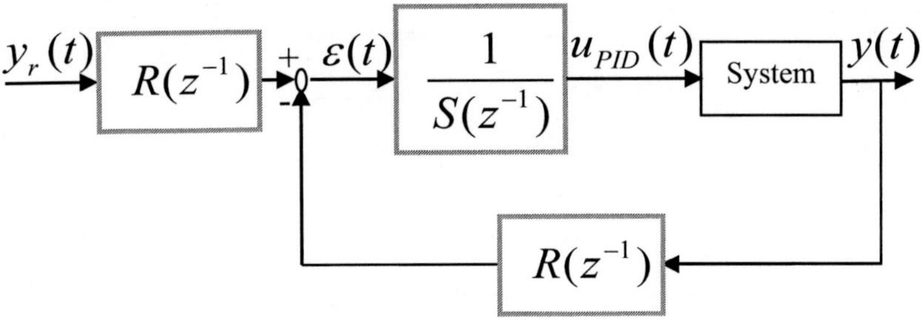

Fig. 5.14. The control system using a proportional-integral-derivative

Given the characteristic of all gains (K_P, K_I, K_D), it is desirable to have a controller that will make the system stable and still produce fast responses and has some robustness properties.

Example: The following figure shows the output signal to a unit step in the command signal for a system with proportional-integral-derivative control.

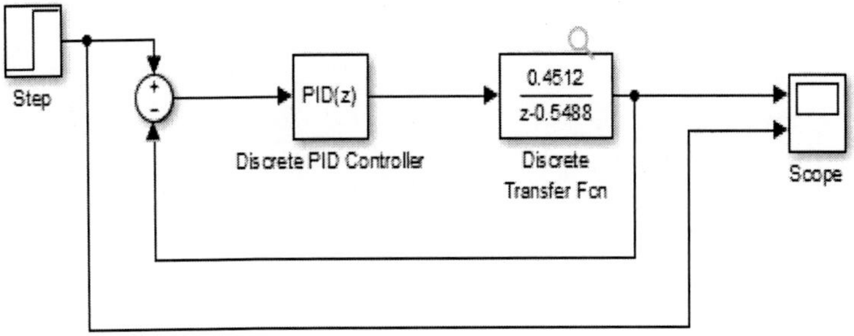

Fig. 5.15. The discrete-time PID controller by Simulink/MatLab

In the next figure, we are going to take three cases of PID gains. In fact, we are going to use $K_P = 1$, $K_D = 1$ and $K_I = 1$ for the first case, $K_P = 0.2$ and $K_I = 0.5$ for the second one however in the last the parameters are: $K_P = 1$, $K_D = 1$ and $K_I = 1$.

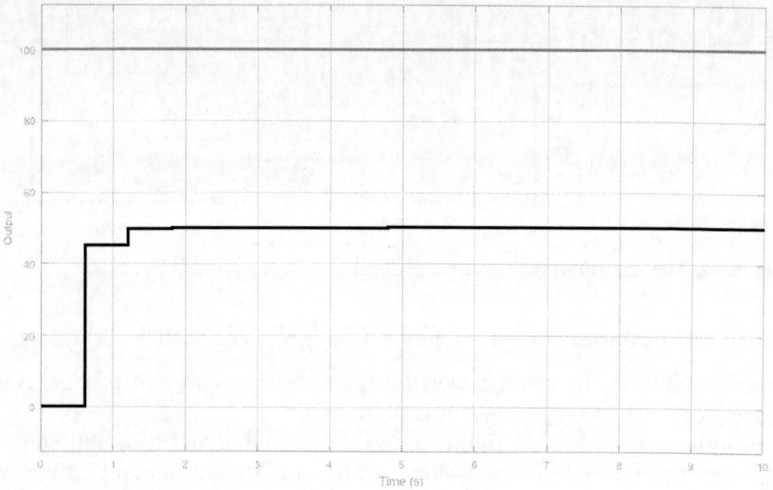

a) $K_P = 1$, $K_D = 1$ and $K_I = 1$

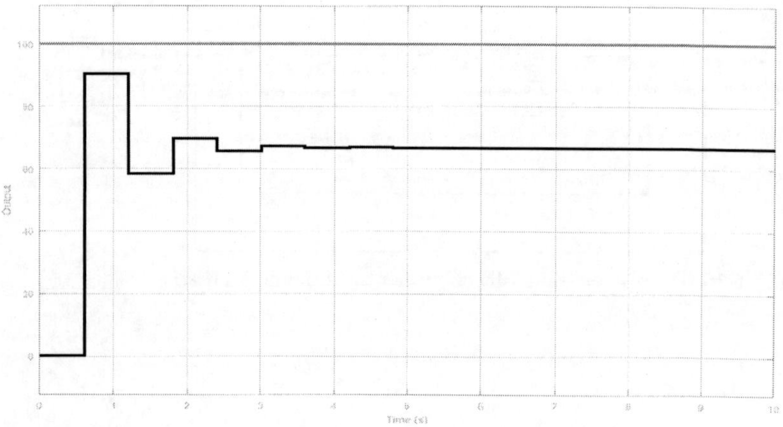

b) $K_P = 0.2$ and $K_I = 0.5$

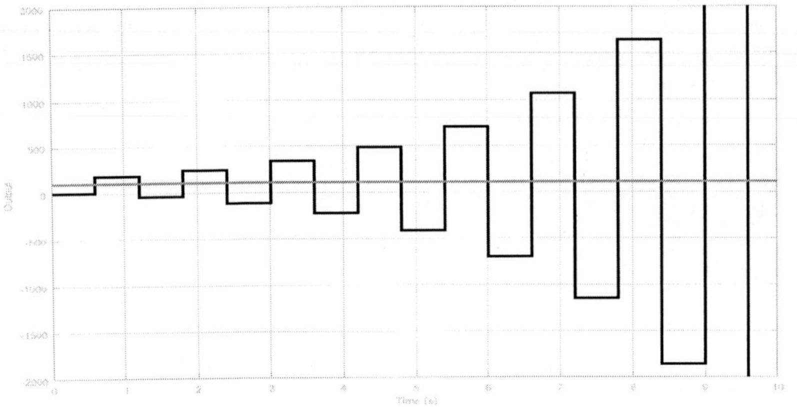

c) $K_P = 1$, $K_D = 1$ and $K_I = 1$

Fig. 5.16. Responses (Black) to step (Blue) changes in the command signal for proportional integral controller.

From this example, proportional-integral-derivative controller has the optimum control dynamics including zero steady state error, fast response (short rise time), no oscillations and higher stability. The necessity of using a derivative gain component in addition to the proportional integral controller is to eliminate the overshoot and the oscillations occurring in the output response of the system.

5.4 Tuning PID parameters method

Tuning a control loop is arranging the control parameters to their optimum values in order to obtain desired control response. At this point, stability is the main necessity, but beyond that, different systems leads to different behaviors and requirements and these might not be compatible with each other. In principle, PID tuning seems completely easy, consisting of only 3 parameters, however, in practice; it is a difficult task because the complex criteria at the PID limit should be satisfied. PID tuning is mostly a heuristic concept but existence of many objectives to be met such as short transient, high stability, rise time and settling time make this process harder.

For these reasons many tuning methods are used. Besides the manual tuning method, different methods for tuning PID parameters have been developed. Some notable techniques are the Ziegler-Nichols method,

Cohen-Coon method, Astrom methods, loop-shapes, Astrom's method, PID tuning software methods (e.g. MATLAB), etc.

5.4.1 PID tuning software method

This method is based on the software technique. In this section we focus on the Simulink/MATLAB example.

Example

The results above show that PID controller for first order system requires tuning. Using the automatic tuning option of MatLab-Simulink one can get the following results:

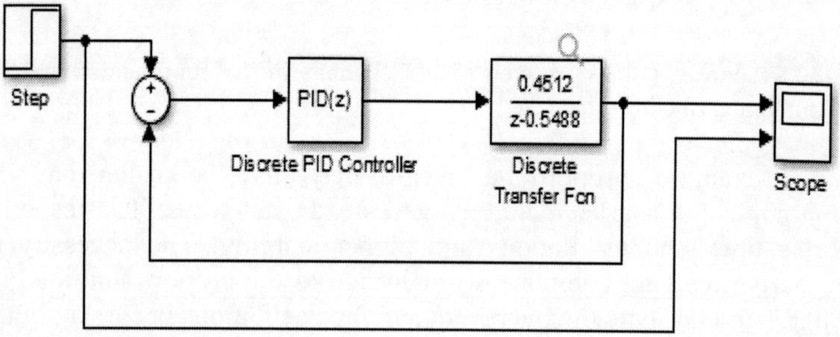

Fig. 5.17. The closed-loop of system using a PID controller by Simulink/Matlab

By using different parameters given by the Table 5.2, the responses of the the system to step changes in the command signal for proportional is given in the Fig. 5.18.

Table 5.2. Controller parameters

Controller Parameters

	Tuned
P	0.013325
I	0.044416
D	0
N	100

Performance and Robustness

	Tuned
Rise time	46.2 seconds
Settling time	84 seconds
Overshoot	0 %
Peak	0.998
Gain margin	37.5 dB @ 1.65 rad/s
Phase margin	86.6 deg @ 0.0444 rad/s
Closed-loop stability	Stable

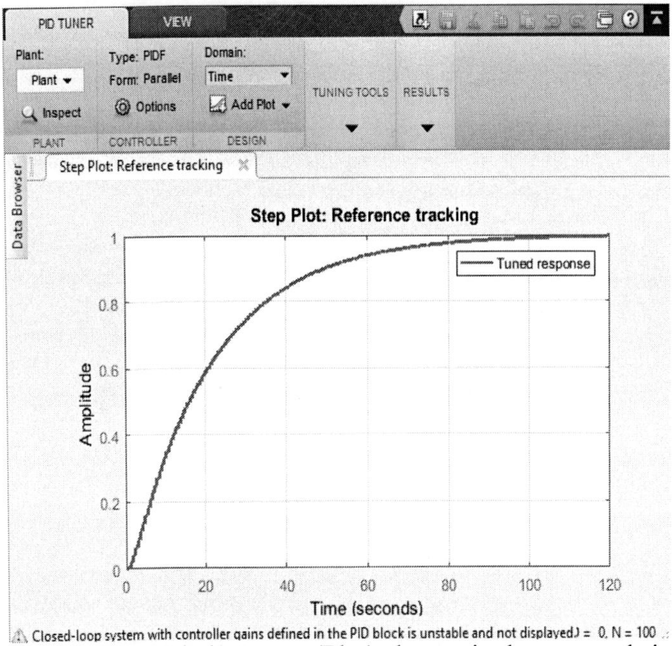

Fig. 5.18. Responses (Black) to step (Blue) changes in the command signal for proportional

5.4.2 Manual tuning method

Manual tuning is achieved by arranging the parameters according to the system response. Until the desired system response is obtained, K_P, K_I and K_D are changed by observing system behavior by lowering and increasing these parameters. Although this method seems simple but it requires a lot of time and experiences.

5.4.3 Pole placement method

This method consists in synthesizing a corrector so that the poles of the closed-loop transfer function are in well-determined positions inside the unit circle. The positions of the poles reflect the performances. In order to apply the pole placement method, some steps are needed:

1- Find the transfer function of the system to be controlled:

$$H(z) = (1 - z^{-1})Z\left[\frac{H(s)}{s}\right] = \frac{b_1 z^{-1} + b_2 z^{-2}}{1 + a_1 z^{-1} + a_2 z^{-2}} = \frac{N(z^{-1})}{D(z^{-1})}$$

2- Find the transfer function in the closed loop

$$H_{CL}(z) = \frac{Y(z^{-1})}{Y_r(z^{-1})} = \frac{N(z^{-1})R(z^{-1})}{D(z^{-1})S(z^{-1}) + N(z^{-1})R(z^{-1})} = \frac{N_{CL}(z^{-1})}{D_{CL}(z^{-1})}$$

with the characteristic equation is

$$D_{CL}(z^{-1}) = 1 + p_1 z^{-1} + p_2 z^{-2}$$

3- Compute controller parameters solving the characteristic equation:

$$D_{CL}(z^{-1}) = D(z^{-1})S(z^{-1}) + N(z^{-1})R(z^{-1}) = 1 + p_1 z^{-1} + p_2 z^{-2}$$

we find:

$$D_{CL}(z^{-1}) = (1 + a_1 z^{-1} + a_2 z^{-2})(1 - z^{-1})(1 - pz^{-1}) + (b_1 z^{-1} + b_2 z^{-2})(r_0 + r_1 z^{-1} + r_2 z^{-2})$$

or

$$D_{CL}(z^{-1}) = 1 + (a_1 - p - 1 + b_1 r_0)z^{-1} + (p - a_1 p - a_1 + a_2 + b_1 r_1 + b_2 r_0)z^{-2} +$$
$$(a_1 p - a_2 p - a_2 + b_1 r_2 + b_2 r_1)z^{-3} + (a_2 p + b_2 r_2)z^{-4} = 1 + p_1 z^{-1} + p_2 z^{-2}$$

By identification we have:

$$\begin{cases} p_1 = a_1 - p - 1 + b_1 r_0 \\ p_2 = p - a_1 p - a_1 + a_2 + b_1 r_1 + b_2 r_0 \\ 0 = a_1 p - a_2 p - a_2 + b_1 r_2 + b_2 r_1 \\ 0 = a_2 p + b_2 r_2 \end{cases}$$

Finally, the controller parameters are:

$$\begin{cases} K_P = -\dfrac{r_0 p + r_1 + (2-p) r_2}{(1-p)^2} \\ T_I = T_s \dfrac{K_P(1-p)}{r_0 + r_1 + r_2} \\ T_D = T_s \dfrac{p^2 r_0 + p r_1 + r_2}{K_P (1-p)^3} \\ \dfrac{T_D}{N} = \dfrac{p T_s}{1-p} \end{cases}$$

Remarks on higher order systems: When the given system has a dynamic order higher than 1 and/or a general PID controller is used, the overall closed-loop transfer function from Y_r to Y will have an order larger than 2, e.g.,

$$H(z) = \frac{b_0 + b_1 z + \ldots + b_m z^m}{a_0 + a_1 z + \ldots + a_n z^n}$$

with $a_n = 1$ and $m \leq n$, $n > 2$.

In this case, we should place the poles of the above transfer function by comparing it to the following desired transfer function

$$H_{desired}(z) = \frac{N(z)}{(z-\alpha_1)\ldots(z-\alpha_p)(z-z_p)(z-\overline{z}_p)}$$

i.e., by placing all the rest poles close to the origin, which is the fastest location in digital control. Eventually, dynamics associated with the poles close to the origin will die out very fast and the overall system is dominated by the pair left. This is left for students to practice in tutorial questions.

5.4.4 Ziegler-Nichols method

More than six decades ago, P-I controllers were more widely used than P-I-D controllers. Despite the fact that P-I-D controller is faster and has no oscillation, it tends to be unstable in the condition of even small changes in the input set point or any disturbances to the process than P-I controllers.

Ziegler-Nichols Method is one of the most effective methods that increase the usage of P-I-D controllers. However, this Cohen-Coon method has been discovered almost after a decade than the Ziegler-Nichols method and it can only be used for first order systems.

Advantages:

- It is an easy experiment; only need to change the P controller
- Includes dynamics of whole process, which gives a more accurate picture of how the system is behaving
- Ziegler-Nichols can be used for any order of the systems, especially for the higher ones,

Disadvantages:

- Experiment can be time consuming
- It can venture into unstable regions while testing the P controller, which could cause the system to become out of control
- For some cases it might result in aggressive gain and overshoot

5.5 Practical application of digital PID controller for DC motor speed

5.5.1 Introduction

In this sub-section, we are focus on to apply a PID controller to obtain the same desired speed in terms of theoretical and practical cases using practical DC motor. Using a PID controller is to make the actual motor speed match the desired motor speed.

We are proposed, using Arduino Uno, a PID algorithm that will calculate necessary power changes to get the actual speed. This will create a cycle where the motor' speed is constantly being checked against the desired speed. The power level is always set based on what is needed to achieve the correct results.

By using PID controller, we can make the steady state error zero with integral control. We can also obtain fast response time by changing the PID parameters. PID is also very feasible when it is compared with other controllers.

5.5.2 The design requirements of the system

The design requirements of the systems may vary from one system to another. For our case, we want a fast response of the system to an error. The overshoot of the system should not be higher than 5% and the settling time should be smaller than 2 seconds.

The main design requirements are as follows;

- Settling time should be less than 2 seconds;
- Overshoot of the system should be less than 5%;
- Steady state error should be less than 1%

5.5.3 Presentation of the DC Motor

The parameters of the DC motor may change according to different torque and revolution per minute "rpm" values of the DC motor. In this application, the used revolution per minute 1000 rpm, the main component and their values are summarized in the next table.

Table 5.3. Different component of DC motor

Component	**Value**
Resistance	$R = 1\Omega$
Rotor moment of inertia	$J = 0.01 kgm^2 s^{-2}$
Inductor	$L = 0.5H$
Motor Viscous Friction Constant	υ=0.1(N.m)/(rad/s)
Electromotive Force Constant	$K = 0.01 NmAmp^{-1}$

5.5.4 The schematic of the DC Motor

The DC motor is modelling by a resistance R, an inductor L and a rotor moment of inertia J as given by the following figure.

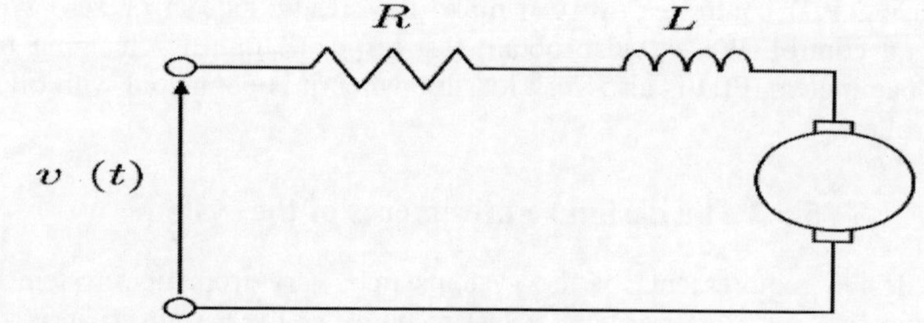

Fig. 5.19. The schematic model of the DC Motor

From this figure, the DC motor is modelling by the two mathematical equations given by the following expression.

$$s\left(Js+b\right)Q(s)=KI(s)$$

$$\left(Ls+R\right)I(s)=V(s)-KQ(s)$$

the transfer function of the DC motor is given by the following expression.

$$\frac{Q(s)}{V(s)}=\frac{K}{\left(Js+b\right)\left(Ls+R\right)+K^{2}}$$

5.5.5 Presentation of the used card

In this application we have used:

- an Arduino uno card (ATMEGA328 microcontroller, 8 bits at 16 MHz, internal power supply 5 V, digital inputs/outputs 14 including 6PWM, analog inputs 6 CAN 10bits, max current per inputs/outputs 40 mA, flash memory 32 KB, SRAM 2KB)
- a DC motor
- a magnet integrated into the engine plate
- a HALL effect sensor
- a transistor for controlling the motor
- 3 LED

The input/output of the Arduino card are presented in the following table:

Table 5.4. Input/Output of the used card

Designation	Input/Output
LED Green	Output D3
LED Orange	Output D5
LED Red	Output D6
DC	Output D9 (PWM)
A Hall effect sensor	Input D2

The schematic model of the DC motor is given by the following figure.

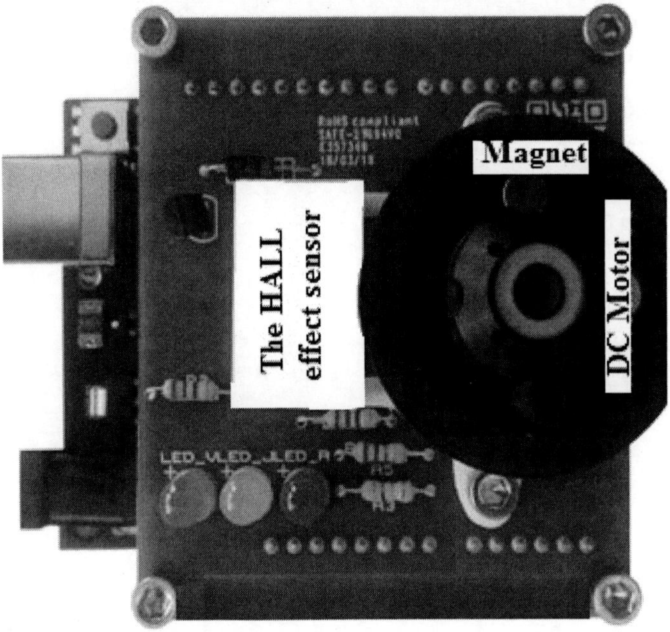

Fig. 5.20. Presentation of the control card

Speed measurement principle: The HALL effect sensor generates a pulse each time the magnet passes. The measurement of the time between two pulses makes it possible to calculate the speed of motor.

5.5.6 Principle of engine identification on the card side

The pseudo-random binary sequence (SBPA) is a random binary signal, of zero mean value, similar to white noise, with an autocorrelation function close to a Dirac, δ (k) = 1.

This sequence is obtained through the use of "N" shift registers looped with an "EXCLUSIVE OR".

PRBS is widely used especially in the identification of systems. It is preferred at the step as an excitation signal. For our case, we used a SBPA of order 10, the loopback is located at the level of bits 7 and 10.

In the lines of the proposed DC motor identification algorithm we will perform, first of all, a generation of excitation by SBPA, then, measurements of the system response (engine speed) and, finally, data transfer to MatLab: excitation and response. The following Arduino card-side DC identification proposed algorithm is summarized in some steps.

The following Arduino card-side DC identification proposed algorithm:

```
# define DC_pin 9
# define HALL_pin 2
# define LED_O 5
# define LED_G 3
# define LED_R 6
```

```
unsigned long time;
unsigned long OldTimes=0;
unsigned long Duration=0;
unsigned long CalculationTime;
unsigned long OldCalculationTime;
boolean rad[10]={0, 1, 1, 1, 1, 1, 1, 1, 1, 1};
boolean sbpa_b[1024]; byte j; int k;
int NominalSpeed=255;
int Ts=15000;
int size=1024;
unsigned long Speed;
float SpeedConversion=62831853071.8;
int cmdPWM;
void setup() {pinMode(DC_pin, OUTPUT);
pinMode(HALL_pin, INPUT);
pinMode(LED_G, OUTPUT);
pinMode(LED_O, OUTPUT);
pinMode(LED_R, OUTPUT);
attachInterrupt(digitalPinToInterrupt(HALL_pin),
TimeCount, FALLING);
Serial.begin(250000);
analogWrite(DC_pin, 0);
digitalWrite(LED_O, HIGH);
delay(5000); digitalWrite(LED_O, LOW);
digitalWrite(LED_R, HIGH);
sbpa_b[0]=rad[0];
cmdPWM=NominalSpeed*sbpa_b[0];
```

5.5.7 Response of the DC motor following the excitation by SBPA signal

Following the excitation of the D motor by the SBPA signal and its response given by the Hall effect sensor transferred to Matlab, by the proposed algorithm in the previous paragraph, we can now identify the transfer function of our DC motor by tracing its response using the

```
cmdPWM=NominalSpeed*sbpa_b[0];
analogWrite(DC_pin, cmdPWM);
for (i=1; i<size; i++)
{OldTimesCalcul=micros();
 sbpa_b[i]=1;
 if (rad[6]==rad[9]){sbpa_b[i]=0;}
 cmdPWM=NominalSpeed*sbpa_b[i];
analogWrite(DC_pin, cmdPWM);
 if (Duration==0){Speed=0;}
else {Speed=SpeedConversion/Duration; }
 for (k=0; k<=24; k+=8) {Serial.write(Speed>>k);}
for (k=0; k<=24; k+=8) {Serial.write(sbpa_b[i]>>k);}
 Shift_rad();
 rad[0]=sbpa_b[i];
 CalculationTime=micros()-OldCalculationTime;
 delayMicroseconds(Ts-CalculationTime);}
detachInterrupt(digitalPinToInterrupt(HALL_pin));
analogWrite(DC_pin, 0);
Serial.end();
 digitalWrite(LED_R, LOW);
 digitalWrite(LED_G, HIGH);}
void loop () {}
void TimeCount(){ time=micros();
Duration=time-OldTimes; OldTimes=time;}
 void Shift_rad() {rad[9]=rad[8]; rad[8]=rad[7]; rad[7]=rad[6];
 rad[6]=rad[5]; rad[5]=rad[4]; rad[4]=rad[3]; rad[3]=rad[2];
rad[2]=rad[1]; rad[1]=rad[0]; }
```

By applying the previous proposed algorithm, we called the obtained values using the following Simulink figure.

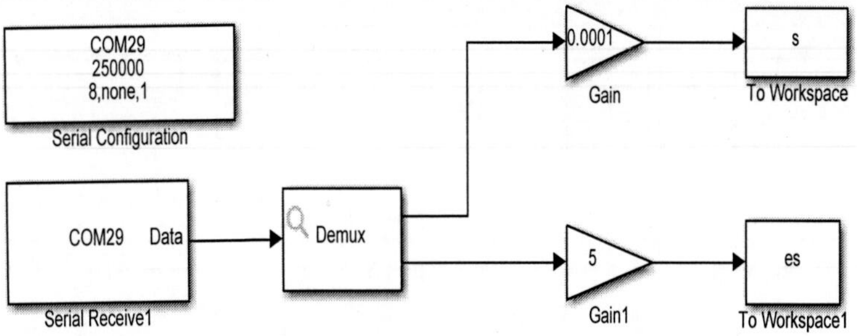

Fig. 5.21. The Simulink model of the DC motor

This figure gives the DC motor speed and the excitation signal 'es' or SBPA signal. Then, these variables are shown in the next figures.

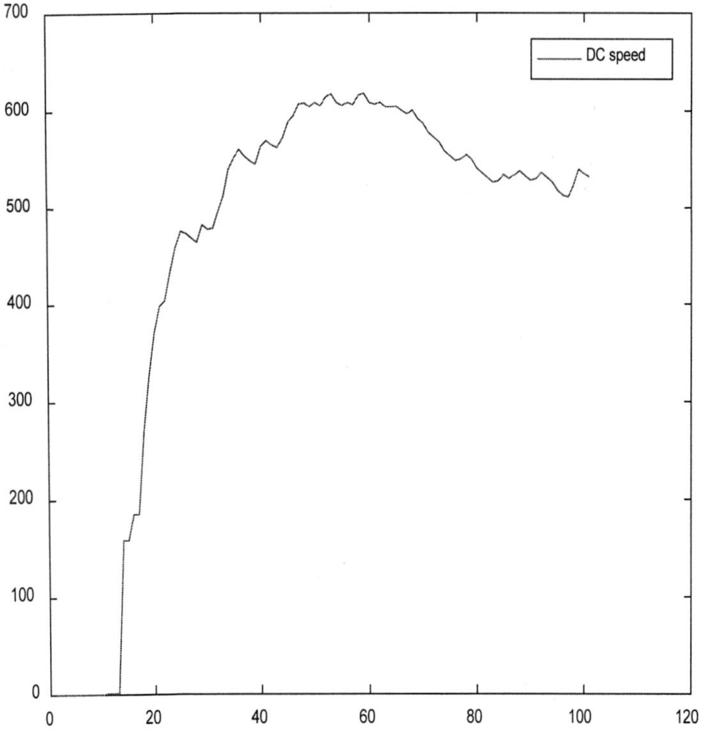

Fig. 5.22. The DC speed

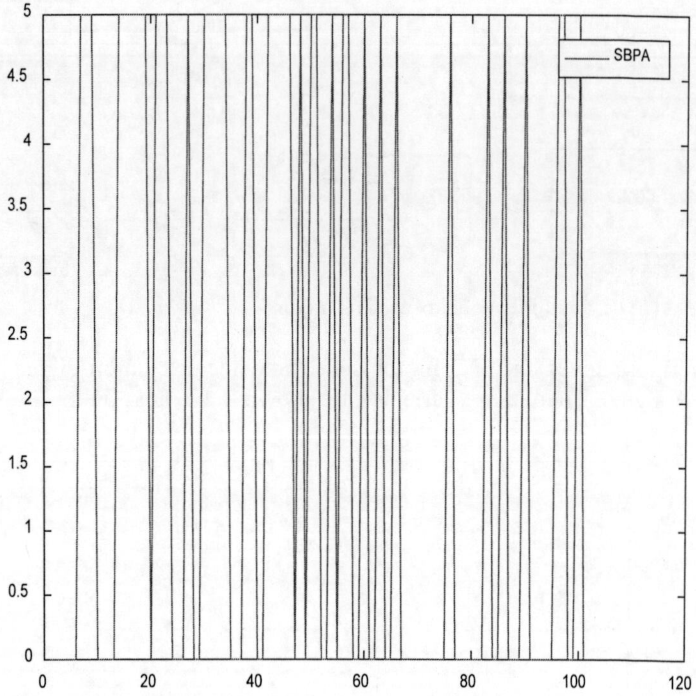

Fig. 5.23. The SBPA signal

5.5.8 Transfer function of DC motor

Using MatLab/Simulink, we present the whole necessary steps in order to find the transfer function of the DC motor, in this section:

1- The first step is to use "system identification" command in MatLab as given in the following figure

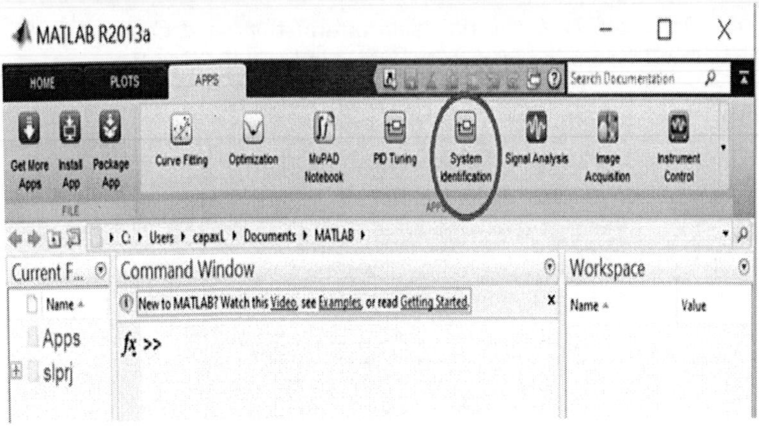

Fig. 5.24. The system identification command

2- The second step is to use "import data" from DC motor which connected with Arduino using the window command in MatLab as given in the following figure

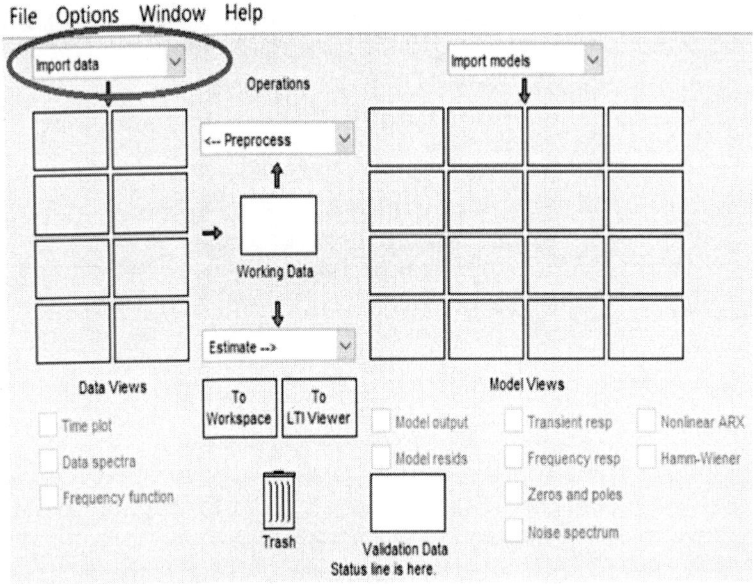

Fig. 5.25. Import data from DC motor

3- The third step is to use the "data information" and choose the "time-domain" and "sampling interval" as given in the following figure

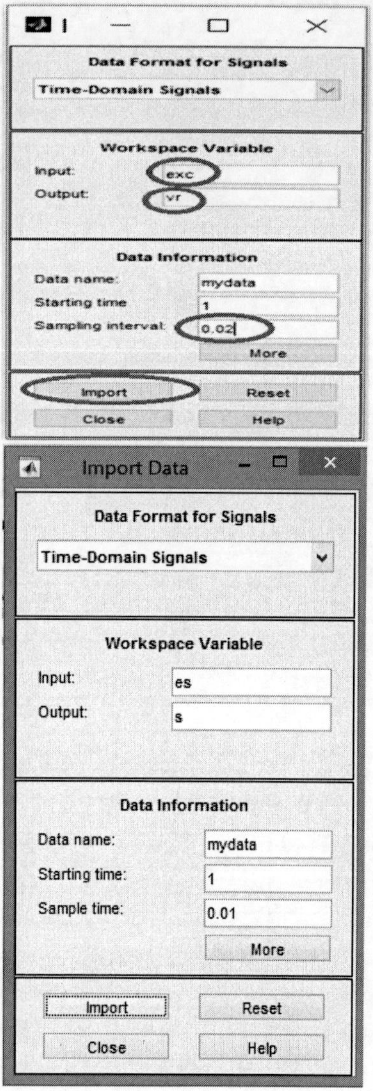

Fig. 5.26. Time-domain and sampling period

4- The fourth step is to estimate the transfer function model of the DC motor by using the corresponding window as given in the following figure

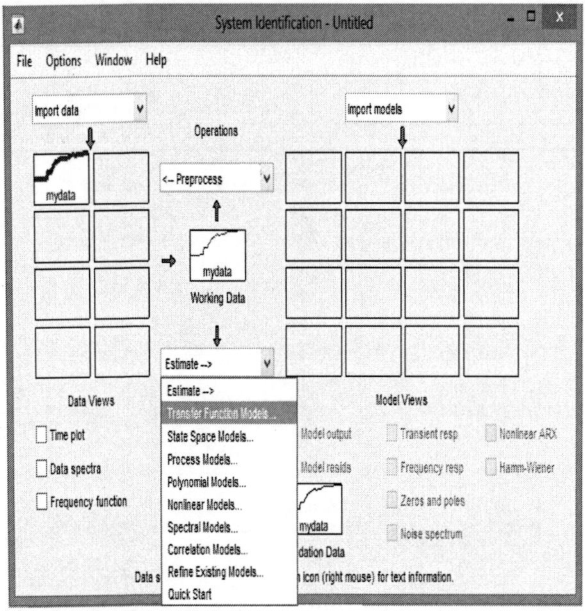

Fig. 5.27. Transfer function estimation

5- The fifth step is to fix the poles, the zeros and the sampling period of the transfer function as given in the following figure

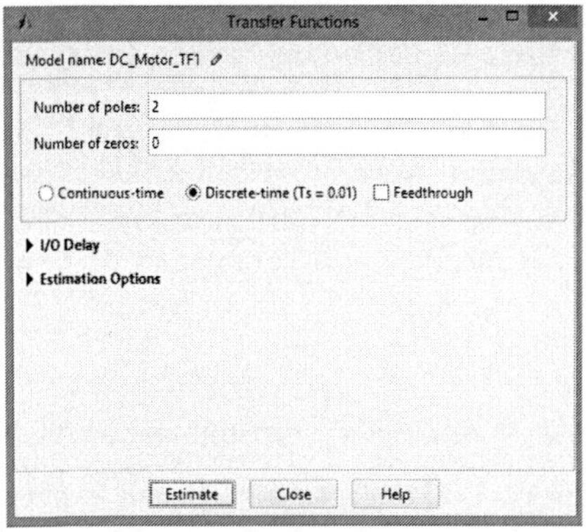

Fig. 5.28. Poles and zeros of the transfer function

6- The sixth step is to move DC_Motor_TF1 into the MatLab workspace environment as given in the following figure

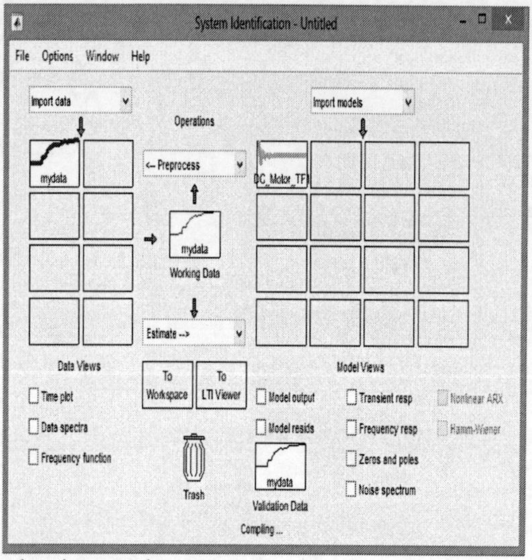

Fig. 5.29. Move data into workspace

Finally, the following equation is the transfer function of the discrete-time DC motor:

$$H(z^{-1}) = \frac{0.048}{z^{-2} - 0.4699z^{-1} - 0.5301}$$

5.5.9 The use of PID controller of the DC motor

In this section it is focused on the parameter of the PID controller. Indeed, it is proposed an algorithm to compare the actual speed of the DC motor with the desired one. The error between theoretical and practical values is corrected with PID controller. The parameters of the PID controller are determined with MATLAB results. Using Simulink/MatLab, some steps are detailed to find the PID parameters.

1- The first step is to use "PID Tuning" in MatLab as given in the following figure

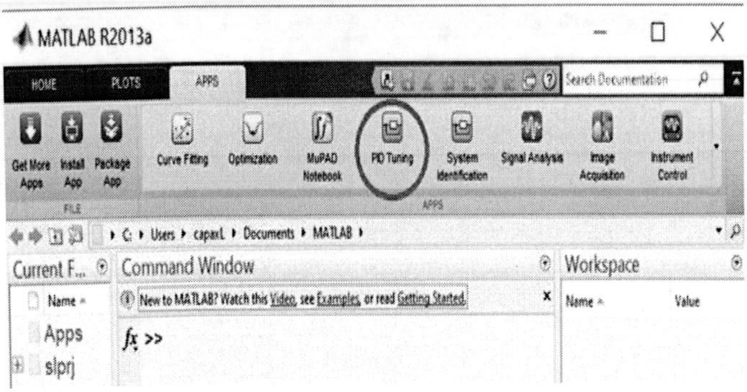

Fig. 5.30. PID tuning

2- The second step is to use "import" given by the window "Import Linear System" in MatLab as given in the following figure.

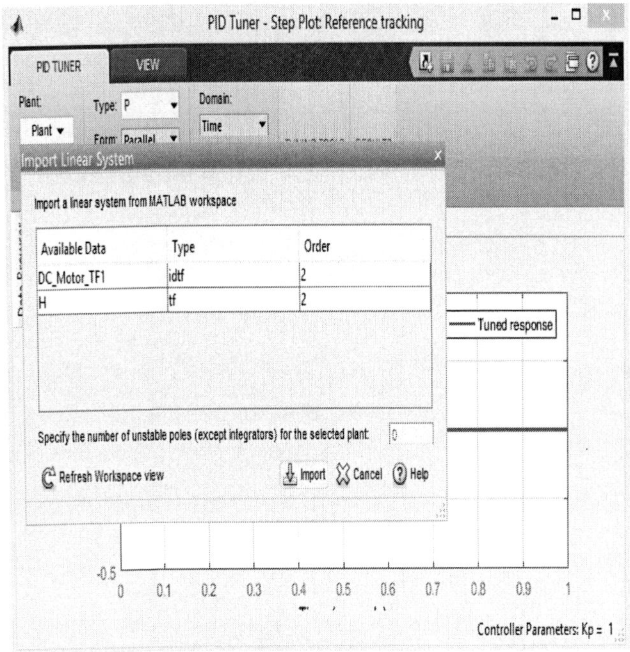

Fig. 5.31. Import data of the DC Transfer function

3- The third step is to plot the DC motor speed as given in the following figure.

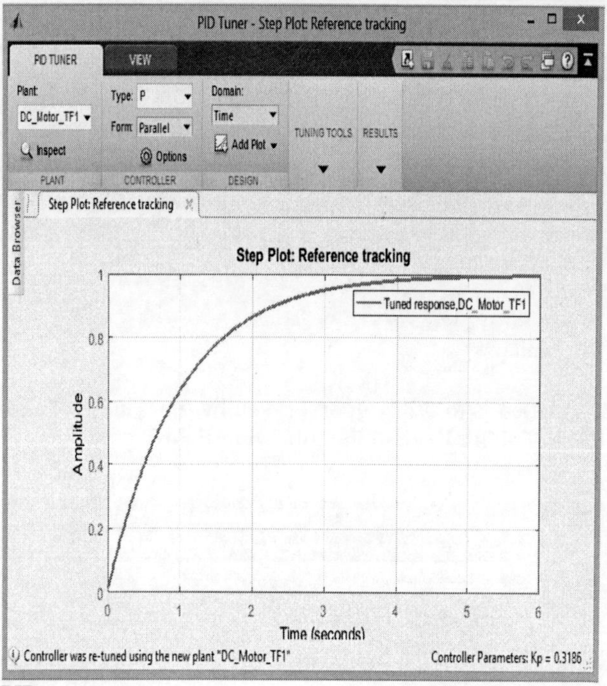

Fig. 5.32. PID tuner

4- The fourth step is to choose the controller type as P, D, I, PD, PI or PID controller. In addition, by using the "show parameters", it is possible to fix the speed of the response as "slower" or "faster" and the robust of the response as "more aggressive" or "more robust" as given by the next figure.

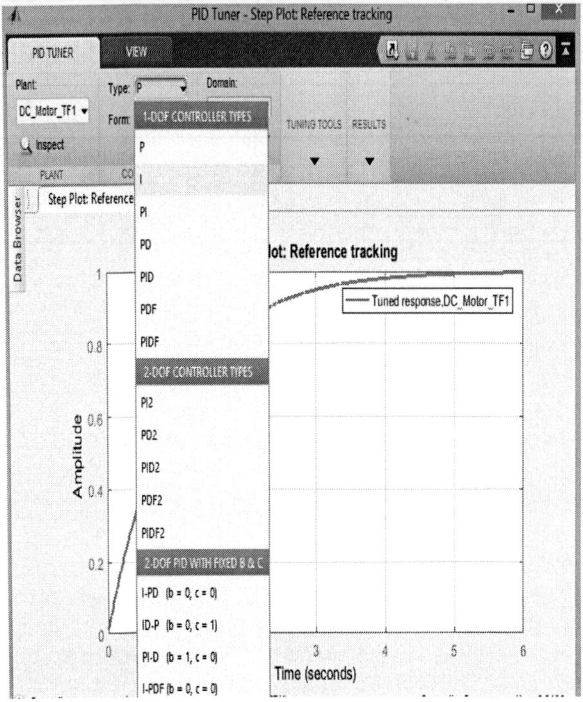

Fig. 5.33. Controller types

5- The fifth step is to choose the used parameters of the controller (K_P , K_I or K_D) from MatLab window and to transfer them to the proposed algorithm using the Arduino card as summarized in the next steps:

```
# define DC_pin 9
# define HALL_pin 2
unsigned long time;
unsigned long OldTime = 0;
unsigned long Duration = 0;
 unsigned long CalculationTime;
 unsigned long OldCalculationTime;
 float deriv_error=0; float integral_error=0;
float kp=0.12173; float ki=--0.0; float kd = -0.0;
```

```
int Ts = 15000; float st= 1500.0; int k; int i; int NbVal=1000;
 float err_p; float err; float ctr_spd; int ctr;
void setup() {Serial.begin (250000);
pinMode(DC_pin, OUTPUT);
 pinMode(HALL_pin, INPUT);
 attachInterrupt(digitalPinToInterrupt(HALL_pin),
TimeCount, FALLING);}
 void loop(){for
(i=1;i<=NbVal;i++){OldCalculationTime=micros();
 if (Duration == 0){spd_mes = 0; }
else {spd_mes = float(60000000/Duration); }
 err =st- spd_mes;
 integral_error = integral_error +err* Ts/1000000;
 ctr_spd = kp*err+ki*integral_error;
if (ctr_spd >= 5.0) {ctr = 255; }
else if (ctr_spd<= 0) {ctr= 0;}
else {ctr= int(ctr_spd*51.0); }
 analogWrite(DC_pin, ctr);
 for (k = 0; k <= 8; k += 8) {int j = spd_mes >> k
; Serial.write(j);}
for (int k = 0; k <= 8; k += 8){int j=ctr>> k; Serial.write(j);}
CalculationTime= micros()-OldCalculationTime;
delayMicroseconds(Ts-CalculationTime);}
analogWrite(DC_pin,0); }
void TimeCount(){time = micros();
Duration=time-OldTime; OldTime=time;}
```

Table 5.5. Controller parameters

Controller Parameters

	Tuned
Kp	0.59179
Ki	0.4744
Kd	n/a
Tf	n/a

Performance and Robustness

	Tuned
Rise time	0.68 seconds
Settling time	4.12 seconds
Overshoot	18.9 %
Peak	1.19
Gain margin	Inf dB @ NaN rad/s
Phase margin	69 deg @ 2 rad/s
Closed-loop stability	Stable

The following figure presents how to use the Simulink block to plot the motor speed and the control law by applying the PID controller.

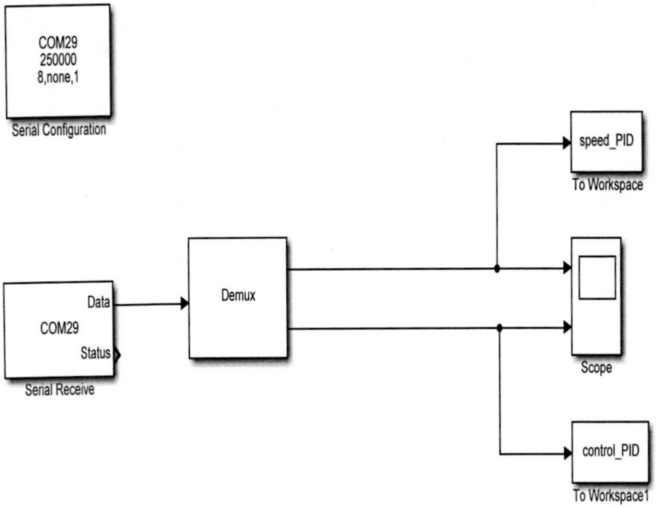

Fig. 5.34. The Simulink block of the motor speed and the control law

5.6 Identification of linear system

The system control needs an appropriate phase is the identification. The aim of system identification is to determine the complexity and the structure of the model. An appropriate method is used to estimate the unknown system model parameters after determining the structure and complexity of the model.

Modelling is the first step in system identification from experimental data. The behavior of the model is determined by the structure of the system and by the properties of the equations describing the relations of the action members.

The way the individual subsystems are interconnected and how they operate is described by the overall system and its behavior.

The behavior of the system obtained using equations describing the physical model can be described in detail by a set of algebraic and differential equations.

5.6.1 System identification background

In system control and signal processing, the mathematical description of the relationship between inputs and outputs of the system represents a dynamic system model. Based on this context, it is possible to determine the system transfer function and thus to identify the system. Basic methods of identification may include methods such as transition and impulse characteristics. The excited input has the character of a single jump or a unit pulse, and the output signal states the model. The application of these techniques is simple, not very susceptible to noise. Another drawback of using these techniques to identify the system is the need to introduce a unit jump / input impulse, which is undesirable for some systems. For this reason, we are also addressing other systems identification approaches that are described in the following text.

The least squares method and the gradient method can be used to estimate the parameters of any linear system.

For simplicity and clarity, consider the transport delay $d = 1$. The following equation states that

$$y(k) = \varphi^T(k-1)\theta$$

where

$\theta = \begin{bmatrix} -a_1 & \dots & -a_n & b_0 & \dots & b_m \end{bmatrix}^T$ and

$\varphi(k-1) = \begin{bmatrix} y(k-1) & \dots & y(k-n) & u(k-1) & \dots & u(k-m) \end{bmatrix}^T$.

θ is the vector of the system parameters and $\phi(k-1)$ is called a regression vector as it is made up of previous system inputs and outputs that affect the current system output value.

When determining the correct system parameter values, it is necessary to determine the initial estimate $\hat{\theta}(0)$. Then the parameter values are so adjusted that the difference between the estimated system output $\hat{y}(k) = \phi(k-1)^T \hat{\theta}(k-1)$ and the actual output of the system $y(k) = \phi(k-1)^T \theta$ is minimized in time. The task of adaptation is thus minimization of the error between the difference of the expected and the actual output.

$$e(k) = \left|y(k) - \hat{y}(k)\right| = \left|\varphi(k-1)^T \theta - \varphi(k-1)^T \hat{\theta}(k-1)\right|$$

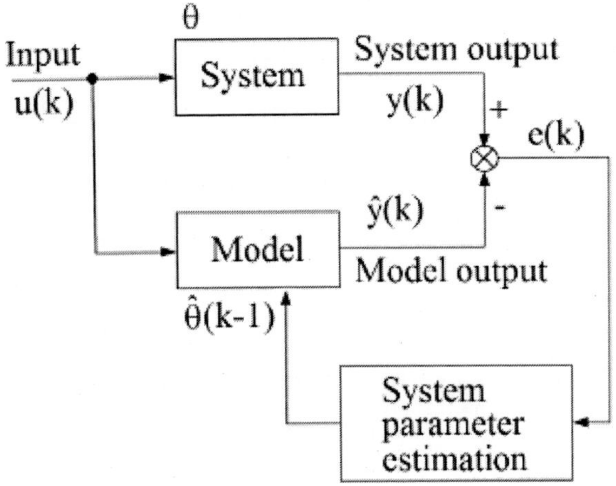

Fig. 5.35. The adaptation scheme of model parameters

5.6.2 System identification using the least squares method

The Adjusted Moving Average AutoRegressive (ARMA) method uses a very universal representation model, translating an input-output relationship of the process of the recurrence type taking into account a possible delay. The ARMA method is used when there are no disturbances.

The adjustment of the coefficients is done by measuring the input-output signals on the actual process for a wide range of input signals.

The behavior model defined in the ARMA method is of the discrete type: this choice facilitates its use in the complex control algorithms which will be established on a digital machine; moreover, the simulation of the results (validation of the model) is direct by digital computation. The model parameters are adjusted by a recursive least squares method, which minimizes the size of the data involved in the calculations. This technique also makes it possible to model systems whose parameters change over time: the model follows these variations with an adjustable "filtering" effect.

The ARMA model: **The ARMA model corresponds to a discrete transfer function developed in the form of a recurrence which establishes a linear relation between the series of inputs $\{u(k)\}$ and that of the modeled outputs $\{y_m(k)\}$ of the form:**

$$y_m(k) = -a_1 y_m(k-1) - \ldots - a_n y_m(k-n) + b_0 u(k) + b_1 u(k-1) + \ldots + b_m u(k-m)$$

The input-output of the system are given by the following scheme:

Fig. 5.36. The ARMA model

Taking into account of a pure delay: If the system has a pure delay T_r (in seconds) corresponding to r sampling periods (r integer), this is taken into account in the model through a translation of all the input data:

$$a_0 y_m(k) + a_1 y_m(k-1) + \ldots + a_n y_m(k-n) = b_0 u(k-r) + b_1 u(k-r-1) + \ldots + b_m u(k-r-m)$$

In the following, the writing of the relations will be done without pure delay which it is easy to take into account in the programs.

Adjusting the parameters of the ARMA model: **The identification of the coefficients of the ARMA model is based on the observation of the signals $u(k)$ and $y(k)$. It should be noted that the shape of the signal $u(k)$ does not intervene directly in the identification phase,**

unlike the methods which rely on a particular excitation of the system (impulse, step, SBPA ...).

5.6.3 The non-recursive least squares method

The most widely used method to determine parameters a_i and b_i of the model is to evaluate the root mean square error between the actual output of the process $y(k)$ and that of the model $y_m(k)$ and to adjust the parameters of the model to minimize this squared error.

The output y_m of the model at time k knowing the sequence of the inputs can be written (assuming that $a_0 = 1$):

$$y_m(k) = -a_1 y_m(k-1) - ... - a_n y_m(k-n) + b_0 u(k) + b_1 u(k-1) + ... + b_m u(k-m)$$

Let $e(k)$ be the difference between the real output $y(k)$ of the process and the output $y_m(k)$ of the model at time k:

$$e(k) = y(k) - y_m(k)$$

or

$$y(k) = e(k) + y_m(k)$$

then

$$y(k) = -a_1 y(k-1) - ... - a_n y(k-n) + b_0 u(k) + b_1 u(k-1) + ... + b_m u(k-m) + e(k)$$

with $e(k)$ is called residual or prediction error. It is the difference between the real output and the predicted output at time k. We define the vector of the parameters:

$$\theta = \begin{bmatrix} a_1 & ... & a_n & b_0 & ... & b_m \end{bmatrix} \in R^{n+m+1}$$

and the observation vector:

$$\varphi = \begin{bmatrix} -y(k-1) & ... & -y(k-n) & u(k) & ... & u(k-m) \end{bmatrix} \in R^{n+m+1}$$

So the output becomes

$$y(k) = \varphi^T(k)\theta + e(k)$$

Let us suppose that we make N successive measurements on the process of the input-output.

We can write $(N-p)$ times the given equation:

$$y(k)=\varphi^T(k)\theta+e(k)$$

so, the set of relations is grouped together in matrix form, $k=1,\ldots,N$:

$$y(1)=\varphi^T(1)\theta+e(1)$$
$$y(2)=\varphi^T(2)\theta+e(2)$$
$$\vdots$$
$$y(N)=\varphi^T(N)\theta+e(N)$$

then

$$Y_N=\begin{bmatrix} y(1) \\ \vdots \\ y(N) \end{bmatrix}; \qquad \phi_N=\begin{bmatrix} \varphi^T(1) \\ \vdots \\ \varphi^T(N) \end{bmatrix}; \qquad E_N=\begin{bmatrix} e(1) \\ \vdots \\ e(N) \end{bmatrix};$$

Finally, all the relationships are grouped together in the form

$$Y_N=\phi_N\theta+E_N$$

The method for estimating the best values of the parameters is the least-squares method, the main steps of which are recalled here:

To estimate θ, put a quadratic criterion J sum of the squares of the prediction errors:

$$J(\theta)=\frac{1}{N}\sum_{i=1}^{N}e_i^2$$
$$=\frac{1}{N}E_N^T E_N$$
$$=\frac{1}{N}[Y_N-\phi_N\theta]^T[Y_N-\phi_N\theta]$$

The minimum of J is obtained by finding the value θ which cancels out the partial derivatives with respect to each of the components of θ is $\frac{\partial J}{\partial\theta}=0$ and the optimum of θ is given by:

$$\hat{\theta}=\left[\phi_N^T\phi_N\right]^{-1}\phi_N^T Y_N$$

The obtained parameter vector makes it possible to define the coefficients of the ARMA model of the process which minimizes the error between the real outputs and that of the model in the sense of least squares.

Example 1: Let consider the first order linear system:

$$y(k) = -0.5y(k-1) + 2u(k)$$

The table below groups together the measurements taken from the inputs-outputs:

k	1	2	3	4	5
$u(k)$	1	-1	0	-1	1
$y(k)$	2.02	-3.05	1.51	-2.76	3.4

Determine, using the 05 measurements, the vector of parameters estimated $\hat{\theta}(5)$ using the non-recursive least squares method, with $P0 = 1000I$.

We propose this code (MatLab), to solve this problem:

```
clear all;
theta= [0 0]';
Y=[2.02 -3.05 1.51 -2.76 3.4]';
U=[1 -1 0 -1 1]';
N=length(Y);
Ts=1;
T=[0:Ts:(N-1)*Ts];
for i = 2:N
H(i,:) = [-Y(i-1) U(i)];
end
theta = inv(H'*H)*H'*Y
```

Using this code (MatLab) the result is as follows

$$\hat{\theta}(7) = \begin{bmatrix} 0.4962 \\ 2.0297 \end{bmatrix}$$

Example 2: Let consider the first order linear system:

$$y(k) = a_1 k + b_1$$

The table below groups together the measurements taken from the inputs-outputs:

k	1	2	3	4	5	6	7
$y(k)$	3.05	5.01	6.98	9.02	10.99	13	15

Determine, using the 07 measurements, the vector of parameters estimated $\hat{\theta}(7)$ using the non-recursive least squares method, with $P0 = 1000I$. We propose this code (MatLab), to solve this problem:

```
clear all;
theta= [0 0]';
Y=[3.05 5.01 6.98 9.02 10.99 13 15]';
U=[1 2 3 4 5 6 7]';
N=length(Y);
Ts=1;
T=[0:Ts:(N-1)*Ts];
for i = 1:N
H(i,:) = [i 1];
end
theta = inv(H'*H)*H'*Y
```

Using this code (MatLab) the result is as follows

$$\hat{\theta}(7) = \begin{bmatrix} 1.9943 \\ 1.0300 \end{bmatrix}$$

Example 3: Let consider the second order linear system:

$$y(k) = -a_1 y(k-1) - a_2 y(k-2) + b_1 u(k-1)$$

The table below groups together the measurements taken from the inputs-outputs:

k	1	2	3	4	5	6	7	8	9	10
$u(k)$	1	1	1	1	1	1	1	1	1	1
$y(k)$	0	0.1	1.8	0.9	1.1	0.95	0.97	0.99	1.02	1.01

Determine, using the 10 measurements, the vector of parameters estimated $\hat{\theta}(10)$ using the non-recursive least squares method, with $P0 = 1000I$. We propose this code (MatLab), to solve this problem:

```
clear all;
clear all;
theta= [0 0 0]';
Y=[0 0.1 1.8 0.9 1.1 0.95 0.97 0.99 1.02 1.01 ]';
U=ones(10,1);
Ts=1;
T=[0:Ts:(length(Y)-1)*Ts];
for i = 3:length(Y)
H(i,:) = [-Y(i-1) -Y(i-2) U(i-1)];
end
theta=inv(H'*H)*H'*Y
```

Using this code (MatLab) the result is as follows

$$\hat{\theta}(10) = \begin{bmatrix} 0.5336 \\ 0.2114 \\ 1.7947 \end{bmatrix}$$

5.6.4 The recursive least squares method

In this case, the parameters are estimated from an input / output pair at each sampling step. So, the estimation can be done in real time when acquiring data or when reading a data file.

The estimation of the vector θ is given by the following equations:

$$\hat{\theta}_k = \hat{\theta}(k-1) + G_k\left(y(k) - \varphi^T(k)\hat{\theta}(k-1)\right)$$

$$P_k = P_{k-1} - \frac{P_{k-1}\varphi(k)\varphi^T(k)P_{k-1}}{1 + \varphi^T(k)P_{k-1}\varphi(k)}$$

$$G_k = \frac{P_{k-1}\varphi(k)}{1 + \varphi^T(k)P_{k-1}\varphi(k)}$$

Example 1: Let consider the first order linear system:

$$H(z) = \frac{y(z)}{u(z)} = \frac{b_0}{1 - a_1 z^{-1}}$$

The table below groups together the measurements taken from the inputs-outputs:

k	1	2	3	4	5	6	7	8
$u(k)$	1	-1	-1	1	1	-1	1	-1
$y(k)$	2.01	-3.02	-0.502	2.251	0.9	-2.46	3.231	-3.601

Determine, using the 08 measurements, the vector of parameters estimated $\hat{\theta}(8)$ using the recursive least squares method, with $P0 = 1000I$.
We propose this code (MatLab), to solve this problem:

```
clear all;
theta= [0 0]';
Y=[2.01 -3.02 -0.502 2.251 0.9 -2.46 3.231 -3.601]';
U=[1 -1 -1 1 1 -1 1 -1]';
N=length(Y);
Ts=1;
T=[0:Ts:(N-1)*Ts];
Pn=1000*eye(size(theta,1));
for i = 3:length(Y)
hn1=[-Y(i-1) U(i)]';
Kn1 = Pn*hn1/(1+hn1'*Pn*hn1);
thetan1=theta + Kn1*(Y(i) - hn1'*theta);
Pn1 = Pn - Kn1*hn1'*Pn;
Pn=Pn1;
theta=thetan1;
end
theta
```

$$\hat{\theta}(8) = \begin{bmatrix} 0.4955 \\ 2.0067 \end{bmatrix}$$

Example 2: Let consider the second order linear system:

$$y(k) = -a_1 y(k-1) - a_2 y(k-2) + b_1 u(k-1)$$

The table below groups together the measurements taken from the inputs-outputs:

k	1	2	3	4	5	6	7	8	9	10
$u(k)$	1	1	1	1	1	1	1	1	1	1
$y(k)$	0	0.1	1.8	0.9	1.1	0.95	0.97	0.99	1.02	1.01

Determine, using the 10 measurements, the vector of parameters estimated $\hat{\theta}(10)$ using the recursive least squares method, with $P0 = 1000I$.

We propose this code (MatLab), to solve this problem:

```
clear all;
Y=[0 0.1 1.8 0.9 1.1 0.95 0.97 0.99 1.02 1.01]';
U=ones(10,1);
Ts=1;
T=[0:Ts:(length(Y)-1)*Ts];
theta= [0 0 0]';
Pn=1000*eye(size(theta,1));

for i = 3:length(Y)
hn1=[-Y(i-1) -Y(i-2) U(i-1)]';
Kn1 = Pn*hn1/(1+hn1'*Pn*hn1);
thetan1=theta + Kn1*(Y(i) - hn1'*theta);
Pn1 = Pn - Kn1*hn1'*Pn;
Pn=Pn1;
theta=thetan1;
end
theta
```

Using this code (MatLab) the result is as follows

$$\hat{\theta}(10) = \begin{bmatrix} 0.5321 \\ 0.2107 \\ 1.7923 \end{bmatrix}$$

Exercise: Let consider the first order system defined by its following difference equation:

$$y(k)+a_1 y(k-1)=b_1 u(k-1)+e(k)$$

or $u(k)$ and $y(k)$ are respectively the input and the output of the system at the instant kT_s (T_s being the sampling period) and $e(k)$ is a noise which represents the set of disturbances acting on the system.
The table below groups together the values of $u(k)$ and $y(k)$ for $k=1,\ldots,8$ with, $\hat{\theta}_0=0$, $P_0=100I$.

k	1	2	3	4	5	6	7	8
$u(k)$	-1	-1	1	0	1	-1	-1	1
$y(k)$	0.1	-1.55	-0.9	1.85	-0.75	1.8	-2.21	-0.62

1. Using the first 07 measurements, determine the vector of estimated parameters $\hat{\theta}(7)$ using the non-recursive least squares method.

2. Repeat the same work using the 08 measurements to determine $\hat{\theta}(8)$.

3. Using the recursive least squares method calculate $\hat{\theta}(8)$ from $\hat{\theta}(7)$ and compare with the results obtained in **2.**

Solution

1. The equation of the system is given by

$$y(k)=-a_1 y(k-1)+b_1 u(k-1)+e(k)$$

$$=\begin{bmatrix}-y(k-1) & u(k-1)\end{bmatrix}\begin{bmatrix}a_1\\ b_1\end{bmatrix}+e(k)$$

We define the vector of the parameters:

$$\theta=\begin{bmatrix}a_1 & b_1\end{bmatrix}^T$$

and the observation vector:

$$\varphi(k)=\begin{bmatrix}-y(k-1) & u(k-1)\end{bmatrix}^T$$

The system equation becomes:

$$y(k)=\varphi^T(k)\theta+e(k)$$

Using the non-recursive least squares method, the vector of estimated parameters $\hat{\theta}(7)$ is:

$$\hat{\theta}_7 = \left[\phi_7^T \phi_7\right]^{-1} \phi_7^T Y_7$$

with

$$Y_7 = \begin{bmatrix} 0.1 \\ -1.55 \\ -0.9 \\ 1.85 \\ -0.75 \\ 1.8 \\ -2.21 \end{bmatrix}; \qquad \phi_7 = \begin{bmatrix} \varphi^T(1) \\ \varphi^T(2) \\ \varphi^T(3) \\ \varphi^T(4) \\ \varphi^T(5) \\ \varphi^T(6) \\ \varphi^T(7) \end{bmatrix} = \begin{bmatrix} 0 & 0 \\ -0.1 & -1 \\ 1.55 & -1 \\ 0.9 & 1 \\ -1.85 & 0 \\ 0.75 & 1 \\ -1.8 & -1 \end{bmatrix};$$

$$\phi_7^T = \begin{bmatrix} 0 & -0.1 & 1.55 & 0.9 & -1.85 & 0.75 & -1.8 \\ 0 & -1 & -1 & 1 & 0 & 1 & -1 \end{bmatrix}$$

$$\phi_7^T \phi_7 = \begin{bmatrix} 10.44 & 2 \\ 2 & 5 \end{bmatrix}$$

$$\left[\phi_7^T \phi_7\right]^{-1} = \frac{1}{48.2} \begin{bmatrix} 5 & -2 \\ -2 & 10.44 \end{bmatrix}$$

$$\phi_7^T Y_7 = \begin{bmatrix} 7.14 \\ 8.31 \end{bmatrix}$$

Finally, the vector of estimated parameters $\hat{\theta}(7)$ is:

$$\hat{\theta}_7 = \frac{1}{48.2} \begin{bmatrix} 5 & -2 \\ -2 & 10.44 \end{bmatrix} \begin{bmatrix} 7.14 \\ 8.31 \end{bmatrix}$$

$$= \begin{bmatrix} 0.395 \\ 1.503 \end{bmatrix}$$

2- Using the non-recursive least squares method, the vector of estimated parameters $\hat{\theta}(8)$ is:

$$\hat{\theta}_8 = \left[\phi_8^T \phi_8\right]^{-1} \phi_8^T Y_8$$

with

$$Y_8 = \begin{bmatrix} 0.1 \\ -1.55 \\ -0.9 \\ 1.85 \\ -0.75 \\ 1.8 \\ -2.21 \\ -0.62 \end{bmatrix}; \qquad \phi_8 = \begin{bmatrix} \varphi^T(1) \\ \varphi^T(2) \\ \varphi^T(3) \\ \varphi^T(4) \\ \varphi^T(5) \\ \varphi^T(6) \\ \varphi^T(7) \\ \varphi^T(8) \end{bmatrix} = \begin{bmatrix} 0 & 0 \\ -0.1 & -1 \\ 1.55 & -1 \\ 0.9 & 1 \\ -1.85 & 0 \\ 0.75 & 1 \\ -1.8 & -1 \\ 2.21 & -1 \end{bmatrix};$$

$$\phi_8^T = \begin{bmatrix} 0 & -0.1 & 1.55 & 0.9 & -1.85 & 0.75 & -1.8 & 2.21 \\ 0 & -1 & -1 & 1 & 0 & 1 & -1 & -1 \end{bmatrix}$$

$$\phi_8^T \phi_8 = \begin{bmatrix} 15.32 & -0.21 \\ -0.21 & 6 \end{bmatrix}$$

$$\left[\phi_8^T \phi_8\right]^{-1} = \frac{1}{91.93} \begin{bmatrix} 6 & 0.21 \\ 0.21 & 15.32 \end{bmatrix}$$

$$\phi_8^T Y_8 = \begin{bmatrix} 5.77 \\ 8.93 \end{bmatrix}$$

Finally, the vector of estimated parameters $\hat{\theta}(8)$ is:

$$\hat{\theta}_8 = \frac{1}{91.93} \begin{bmatrix} 6 & 0.21 \\ 0.21 & 15.32 \end{bmatrix} \begin{bmatrix} 5.77 \\ 8.93 \end{bmatrix}$$

$$= \begin{bmatrix} 0.39 \\ 1.5 \end{bmatrix}$$

3- Using the recursive least squares method, the vector of estimated parameters $\hat{\theta}(8)$ is:

$$\hat{\theta}_8 = \hat{\theta}_7 + G_8\left(y(8) - \varphi^T(8)\hat{\theta}_7\right)$$

with

$$y(8) = -0.62$$

$$\varphi^T(8) = \begin{bmatrix} 2.21 & -1 \end{bmatrix}$$

$$\hat{\theta}_7 = \begin{bmatrix} 0.395 \\ 1.503 \end{bmatrix}$$

$$G_8 = \frac{P_7 \varphi(8)}{1 + \varphi^T(8) P_7 \varphi(8)}$$

$$P_7 = \left[\phi_7^T \phi_7 \right]^{-1}$$

$$= \frac{1}{48.2} \begin{bmatrix} 5 & -2 \\ -2 & 10.44 \end{bmatrix}$$

$$P_7 \varphi(8) = \frac{1}{48.2} \begin{bmatrix} 5 & -2 \\ -2 & 10.44 \end{bmatrix} \begin{bmatrix} 2.21 \\ -1 \end{bmatrix}$$

$$= \begin{bmatrix} 0.27 \\ -0.308 \end{bmatrix}$$

$$\varphi^T(8) P_7 \varphi(8) = \begin{bmatrix} 2.21 & -1 \end{bmatrix} \begin{bmatrix} 0.27 \\ -0.308 \end{bmatrix}$$

$$G_8 = \begin{bmatrix} \dfrac{0.27}{1.904} \\ -\dfrac{0.308}{1.904} \end{bmatrix}$$

$$= \begin{bmatrix} 0.141 \\ -0.161 \end{bmatrix}$$

Finally, the vector of estimated parameters $\hat{\theta}(8)$ is:

$$\hat{\theta}_8 = \begin{bmatrix} 0.395 \\ 1.503 \end{bmatrix} + \begin{bmatrix} 0.141 \\ -0.161 \end{bmatrix} (-0.61 - (2.21 - 1)) \begin{bmatrix} 0.395 \\ 1.503 \end{bmatrix}$$
$$= \begin{bmatrix} 0.3964 \\ 1.5013 \end{bmatrix}$$

5.6.5 The recursive extended least squares method

If the system is disturbed by a sequence of noise, the adopted model is the autoregressive with moving average and exogenous (ARMAX) model:

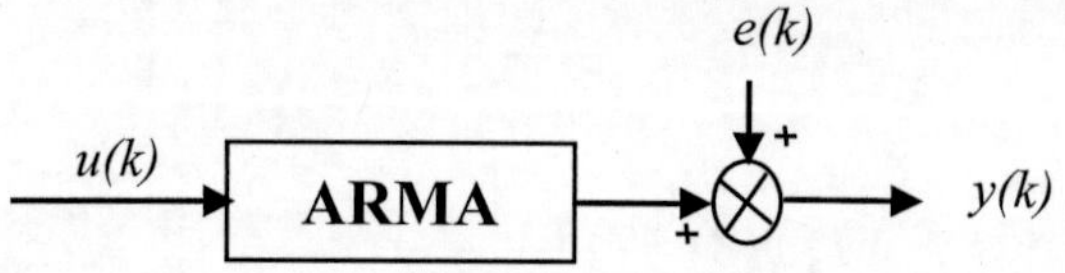

Fig. 5.37. The ARMAX model

The ARMAX model corresponds to a discrete transfer function developed in the form of a recurrence which establishes a linear relation between the series of inputs $\{u(k)\}$, outputs $\{y_m(k)\}$ and disturbances $\{e(k)\}$ of the form:

$$y(k) = -a_1 y(k-1) - ... - a_n y(k-n) + b_0 u(k) + b_1 u(k-1) + ...$$
$$+ b_m u(k-m) + e(k) + c_1 e(k-1) + ... + c_q e(k-q)$$

$e(k)$: disturbances.

The effect of noise is evaluated by the Signal to Noise Ratio (SNR):

$$SNR = \frac{Signal\_Power}{Noise\_Power}$$

Remark: when the SNR is large, the effect of noise on the signal is weak. The recursive extended least squares (RELS) method aims to identify the parameters the parameters of the ARMAX model which are a_i, $i = 1, \cdots, n$, b_j, $j = 0, \cdots, m$ and c_k, $k = 1, \cdots, q$.

The vector of parameters θ:

$$\theta = \begin{bmatrix} a_1 \cdots a_n & b_0 \cdots b_m & c_1 \cdots c_q \end{bmatrix}^T$$

and the observation vector φ:

$$\varphi(k) = \begin{bmatrix} -y(k-1) & \ldots & -y(k-n) & u(k) \cdots u(k-m) & e(k-1) \cdots e(k-q) \end{bmatrix}^T$$

θ and $\varphi(k)$ are the extended vectors to the noise components and the variables $e(k-1),\ldots,e(k-q)$ are not measurable.

The system equation becomes:

$$y(k) = \varphi^T(k)\theta + e(k)$$

In the observation vector, the variables $e(k-1),\ldots,e(k-q)$ are replaced by the prediction errors $\xi(k-1),\ldots,\xi(k-q)$:

$$\varphi(k) = \begin{bmatrix} -y(k-1) & \ldots & -y(k-n) & u(k) \cdots u(k-m) & \xi(k-1) \cdots \xi(k-q) \end{bmatrix}^T$$

The RELS algorithm is written as:

$$\hat{\theta}_k = \hat{\theta}_{k-1} + G_k\left(y(k) - \varphi^T(k)\hat{\theta}_{k-1}\right)$$

$$P_k = P_{k-1} - \frac{P_{k-1}\varphi(k)\varphi^T(k)P_{k-1}}{1 + \varphi^T(k)P_{k-1}\varphi(k)}$$

$$G_k = \frac{P_{k-1}\varphi(k)}{1 + \varphi^T(k)P_{k-1}\varphi(k)}$$

$$\xi(k) = y(k) - \varphi^T(k)\hat{\theta}_{k-1}$$

$\xi(k)$: a priori prediction error.

Example: Let consider the first order linear system:

$$y(k) = a_1 y(k-1) + b_1 u(k) + c_1 e(k-1) + e(k)$$

The table below groups together the values of $u(k)$ and $y(k)$ for $k = 1,\ldots,4$ with, $\hat{\theta}_0 = 0$, $P_0 = 100I$.

k	1	2	3	4
$u(k)$	-1	1	1	-1
$y(k)$	-2.1	1.2	2.7	-0.6

Determine the vector of estimated parameters $\hat{\theta}(3)$ using the recursive extended least squares method.

Solution: we have the given vectors:

$$\theta = \begin{bmatrix} a_1 & b_1 & c_1 \end{bmatrix}^T$$

$$\varphi(k) = \begin{bmatrix} y(k-1) & u(k) & \xi(k-1) \end{bmatrix}^T$$

$$\xi(k) = y(k) - \varphi^T(k)\hat{\theta}_{k-1}$$

Calculation of $\hat{\theta}_1$:

$$\hat{\theta}_1 = \hat{\theta}_0 + G_1\left(y(1) - \varphi^T(1)\hat{\theta}_0\right)$$

or

$$y(1) = -2.1$$

$$\varphi(1) = \begin{bmatrix} y(0) & u(1) & \xi(0) \end{bmatrix}^T$$

$$\xi(0) = y(0) - \varphi^T(0)\hat{\theta}_{-1} = y(0) = 0$$

$$\varphi(1) = \begin{bmatrix} y(0) & u(1) & \xi(0) \end{bmatrix}^T = \begin{bmatrix} 0 & -1 & 0 \end{bmatrix}^T$$

$$G_1 = \frac{P_0\varphi(1)}{1 + \varphi^T(1)P_0\varphi(1)}$$

$$P_0\varphi(1) = \begin{pmatrix} 100 & 0 & 0 \\ 0 & 100 & 0 \\ 0 & 0 & 100 \end{pmatrix}\begin{pmatrix} 0 \\ -1 \\ 0 \end{pmatrix}$$

$$= \begin{pmatrix} 0 \\ -100 \\ 0 \end{pmatrix}$$

$$\varphi^T(1)P_0\varphi(1) = \begin{pmatrix} 0 & -1 & 0 \end{pmatrix}\begin{pmatrix} 0 \\ -100 \\ 0 \end{pmatrix} = 100$$

$$G_1 = \frac{\begin{pmatrix} 0 \\ -100 \\ 0 \end{pmatrix}}{101} = \begin{pmatrix} 0 \\ -\dfrac{100}{101} \\ 0 \end{pmatrix}$$

$$\hat{\theta}_1 = \hat{\theta}_0 + G_1\left(y(1) - \varphi^T(1)\hat{\theta}_0\right) = 0 + \begin{pmatrix} 0 \\ -\dfrac{100}{101} \\ 0 \end{pmatrix}(-2.1 - 0) = \begin{pmatrix} 0 \\ 2.07 \\ 0 \end{pmatrix}$$

Calculation of $\hat{\theta}_2$:

$$\hat{\theta}_2 = \hat{\theta}_1 + G_2\left(y(2) - \varphi^T(2)\hat{\theta}_1\right)$$

or

$$y(2) = 1.2$$

$$\xi(1) = y(1) - \varphi^T(1)\hat{\theta}_0 = y(1) = -2.1$$

$$\varphi(2) = \begin{bmatrix} y(1) & u(2) & \xi(1) \end{bmatrix}^T = \begin{bmatrix} -2.1 & 1 & -2.1 \end{bmatrix}^T$$

$$G_2 = \frac{P_1\varphi(2)}{1 + \varphi^T(2)P_1\varphi(2)}$$

$$P_1 = P_0 - \frac{P_0\varphi(1)\varphi^T(1)P_0}{1 + \varphi^T(1)P_0\varphi(1)}$$

$$P_1 = \begin{pmatrix} 100 & 0 & 0 \\ 0 & 100 & 0 \\ 0 & 0 & 100 \end{pmatrix} - \begin{pmatrix} 0 \\ -\dfrac{100}{101} \\ 0 \end{pmatrix}\begin{pmatrix} 0 & -100 & 0 \end{pmatrix} = \begin{pmatrix} 100 & 0 & 0 \\ 0 & 0.99 & 0 \\ 0 & 0 & 100 \end{pmatrix}$$

$$P_1\varphi(2) = \begin{pmatrix} -210 \\ 0.99 \\ -210 \end{pmatrix}$$

$$\varphi^T(2)P_1\varphi(2) = \begin{pmatrix} -2.1 & 1 & -2.1 \end{pmatrix}\begin{pmatrix} -210 \\ 0.99 \\ -210 \end{pmatrix} = 882.99$$

$$G_2 = \frac{\begin{pmatrix} -210 \\ 0.99 \\ -210 \end{pmatrix}}{883.99} = \begin{pmatrix} 0.237 \\ 0.001 \\ 0.237 \end{pmatrix}$$

$$\xi(2) = y(2) - \varphi^T(2)\hat{\theta}_1 = 1.2 - \begin{bmatrix} -2.1 & 1 & -2.1 \end{bmatrix}\begin{pmatrix} 0 \\ 2.07 \\ 0 \end{pmatrix} = -0.87$$

$$\hat{\theta}_2 = \hat{\theta}_1 + G_2\left(y(2) - \varphi^T(2)\hat{\theta}_1\right) = \begin{pmatrix} 0 \\ 2.07 \\ 0 \end{pmatrix} + \begin{pmatrix} -0.237 \\ 0.001 \\ -0.237 \end{pmatrix}(-0.87) = \begin{pmatrix} 0.206 \\ 2.0697 \\ 0.206 \end{pmatrix}$$

Calculation of $\hat{\theta}_3$:

$$\hat{\theta}_3 = \hat{\theta}_2 + G_3\left(y(3) - \varphi^T(3)\hat{\theta}_2\right)$$

or

$$y(3) = 2.7$$

$$\xi(2) = y(2) - \varphi^T(2)\hat{\theta}_1 = 2.7 - \begin{bmatrix} -2.1 & 1 & -2.1 \end{bmatrix}\begin{pmatrix} 0 \\ 2.07 \\ 0 \end{pmatrix} = 1.2 - 2.07 = -0.87$$

$$\varphi(3) = \begin{bmatrix} y(2) & u(3) & \xi(2) \end{bmatrix}^T = \begin{bmatrix} 1.2 & 2.7 & -0.87 \end{bmatrix}^T$$

$$G_3 = \frac{P_2\varphi(3)}{1+\varphi^T(3)P_2\varphi(3)}$$

$$P_2 = P_1 - \frac{P_1\varphi(2)\varphi^T(2)P_1}{1+\varphi^T(2)P_1\varphi(2)}$$

$$P_1\varphi(2) = \begin{pmatrix} -210 \\ 0.99 \\ -210 \end{pmatrix}$$

$$\varphi^T(2)P_1\varphi(2) = \begin{pmatrix} -2.1 & 1 & -2.1 \end{pmatrix}\begin{pmatrix} -210 \\ 0.99 \\ -210 \end{pmatrix} = 882.99$$

$$P_2 = P_1 - \frac{P_1\varphi(2)\varphi^T(2)P_1}{883.99}$$

$$P_1\varphi(2)\varphi^T(2)P_1 = \begin{pmatrix} 100 & 0 & 0 \\ 0 & 0.99 & 0 \\ 0 & 0 & 100 \end{pmatrix}\begin{bmatrix} -2.1 \\ 1 \\ -2.1 \end{bmatrix}\begin{pmatrix} -2.1 & 1 & -2.1 \end{pmatrix}\begin{pmatrix} 100 & 0 & 0 \\ 0 & 0.99 & 0 \\ 0 & 0 & 100 \end{pmatrix}$$

$$= \begin{pmatrix} 44100 & -207.9 & 44100 \\ -207.9 & 0.9801 & -207.9 \\ 44100 & -207.9 & 44100 \end{pmatrix}$$

$$P_2 = \begin{pmatrix} 100 & 0 & 0 \\ 0 & 0.99 & 0 \\ 0 & 0 & 100 \end{pmatrix} - \frac{1}{883.99}\begin{pmatrix} 44100 & -207.9 & 44100 \\ -207.9 & 0.9801 & -207.9 \\ 44100 & -207.9 & 44100 \end{pmatrix} = \begin{pmatrix} 50.113 & 0.235 & -49.887 \\ 0.235 & 0.989 & 0.235 \\ -49.887 & 0.235 & 50.113 \end{pmatrix}$$

$$P_2\varphi(3) = \begin{pmatrix} 50.113 & 0.235 & -49.887 \\ 0.235 & 0.989 & 0.235 \\ -49.887 & 0.235 & 50.113 \end{pmatrix}\begin{pmatrix} 1.2 \\ 2.7 \\ -0.87 \end{pmatrix} = \begin{pmatrix} 103.77 \\ 1.066 \\ -103.227 \end{pmatrix}$$

$$\varphi^T(3)P_2\varphi(3)=\begin{pmatrix}1.2 & 2.7 & -0.87\end{pmatrix}\begin{pmatrix}103.77\\1.066\\-103.227\end{pmatrix}=215.39$$

$$G_3=\frac{\begin{pmatrix}103.77\\1.066\\-103.227\end{pmatrix}}{216.39}=\begin{pmatrix}0.479\\0.005\\-0.477\end{pmatrix}$$

$$\xi(3)=y(3)-\varphi^T(3)\hat{\theta}_2=2.7-\begin{pmatrix}1.2 & 1 & -0.87\end{pmatrix}\begin{pmatrix}0.206\\2.0697\\0.206\end{pmatrix}=0.56$$

$$\hat{\theta}_3=\hat{\theta}_2+G_3\left(y(3)-\varphi^T(3)\hat{\theta}_2\right)=\begin{pmatrix}0.206\\2.0697\\0.206\end{pmatrix}+\begin{pmatrix}0.479\\0.005\\-0.477\end{pmatrix}0.56=\begin{pmatrix}0.47\\2.18\\-0.061\end{pmatrix}$$

5.7 Applications

Application 1 - Proportional controller

The discrete-time system is given by its transfer function:

$$H(z)=\frac{0.15}{z-0.85}$$

1. Synthesis a proportional controller in which the pole of the closed loop transfer function is $z=0.75$.
2. Synthesis a controller in that the difference equation in the closed loop transfer function is

$$y(k)=0.5r(k-1)+0.3y_r(k-2)+0.2y_r(k-3)$$

with $y_r(k)$ is the reference signal and $y(k)$ is the output answer.

Answer

1. The transfer function of the discrete-time system in open loop is

$$H_{OL}(z) = K_P \frac{0.15}{z-0.85}$$

in closed loop the transfer function is

$$H_{CL}(z) = \frac{H_{OL}(z)}{1+H_{OL}(z)} = \frac{K_P \frac{0.15}{z-0.85}}{1+K_P \frac{0.15}{z-0.85}} = \frac{0.15K_P}{z-0.85+0.15K_P} = \frac{0.15K_P}{z-z_0}$$

with $z_0 = 0.85 - 0.15K_P = 0.75 \Rightarrow \mathrm{K}_P = \frac{0.85\text{-}0.75}{0.15} = 0.6667$

2. In closed loop the transfer function is

$$H_{CL}(z) = \frac{0.15K(z)}{z-0.85+0.15K(z)} = \frac{Y(z)}{Y_r(z)}$$

or the output answer is

$$y(k) = 0.5y_r(k-1) + 0.3y_r(k-2) + 0.2y_r(k-3)$$

the z-transform of the output answer is

$$Y(z) = 0.5z^{-1}Y_r(z) + 0.3z^{-2}Y_r(z) + 0.2z^{-3}Y_r(z) = (0.5z^{-1} + 0.3z^{-2} + 0.2z^{-3})Y_r(z)$$

and the transfer function is

$$\frac{Y(z)}{Y_r(z)} = 0.5z^{-1} + 0.3z^{-2} + 0.2z^{-3}$$

by comparing the two transfer functions

$$\frac{0.15K(z)}{z-0.85+0.15K(z)} = 0.5z^{-1} + 0.3z^{-2} + 0.2z^{-3}$$

so, the expression of the controller is

$$K(z) = \frac{0.5 - 0.125z^{-1} - 0.225z^{-2} - 0.17z^{-3}}{0.15 - 0.075z^{-1} - 0.3z^{-2} - 0.03z^{-3}}$$

Application 2 - Derivative controller

The aim of this application is to compare the simple derivative output controller with a derivative output controller with filter.

We give $T_D = 10T_s$, $N = 8$, $K_P = 1$ and a unit step input with amplitude E_0.

1 - Compute the output of the simple derivative controller.

2 - Prove that the transfer function of a derivative controller is

$$U_D(z) = \frac{K_P T_D}{T_s + \frac{T_D}{N}} \frac{z-1}{z-z_0}$$

with z_0 is to be found.

Compute the output of this simple controller.

3 - What kind of problems can the pole z_0 create?

Answer

Using a unit input step with E_0, $T_D = 10T_s$, $N = 8$ and $K_P = 1$.

1 - The transfer function is

$$C_D(z) = \frac{U_D(z)}{E_D(z)} = \frac{K_P T_D(1-z^{-1})}{T_s} = \frac{K_P 10T_s(1-z^{-1})}{T_s} = 10(1-z^{-1})$$

$$\Rightarrow U_D(z) = 10(1-z^{-1})E_D(z) = 10\frac{z-1}{z}\frac{E_0 z}{z-1} = 10E_0$$

$$\Rightarrow u_D(k) = 10E_0 e(k) \quad ; \quad e(k) = 1 \quad \forall k \in [1, +\infty[$$

2 - The transfer function is

$$C_D(z) = K_P \frac{T_D \frac{(1-z^{-1})}{T_s}}{1 + \frac{T_D}{N}\frac{(1-z^{-1})}{T_s}} = K_P \frac{T_D \frac{(z-1)}{T_s z}}{1 + \frac{T_D}{N}\frac{(z-1)}{T_s z}} = K_P \frac{T_D(z-1)}{T_s z + \frac{T_D}{N}(z-1)}$$

$$C_D(z) = K_P \frac{T_D(z-1)}{z(T_s + \frac{T_D}{N}) - \frac{T_D}{N}} = \frac{K_P T_D}{T_s + \frac{T_D}{N}} \frac{z-1}{z - \frac{\frac{T_D}{N}}{T_s + \frac{T_D}{N}}} = \frac{K_P T_D}{T_s + \frac{T_D}{N}} \frac{z-1}{z-z_0}$$

with $z_0 = \dfrac{\dfrac{T_D}{N}}{T_s + \dfrac{T_D}{N}}$.

$$C_D(z) = \frac{U_D(z)}{E_D(z)}$$

$$= \frac{K_P T_D}{T_s + \dfrac{T_D}{N}} \frac{z-1}{z-z_0}$$

$$= \frac{10T_s}{T_s + \dfrac{10T_s}{N}} \frac{z-1}{z-z_0}$$

$$= \frac{10}{1+\dfrac{10}{8}} \frac{z-1}{z-z_0}$$

$$= 4.4444 \frac{z-1}{z-z_0}$$

$$U_D(z) = 4.4444 \frac{z-1}{z-z_0} E_D(z)$$

$$= 4.4444 \frac{z-1}{z-z_0} \frac{E_0 z}{z-1}$$

$$= 4.4444 E_0 \frac{z^2 - z}{z^2 - z(1+z_0) + z_0}$$

$$U_D(z) = 4.4444 E_0 \frac{1-z^{-1}}{1-(1+z_0)z^{-1} + z_0 z^{-2}}$$

$$\Rightarrow (1-(1+z_0)z^{-1} + z_0 z^{-2}) U_D(z) = 4.4444 E_0 (1-z^{-1})$$

$$u_D(k) - (1+z_0) u_D(k-1) + z_0 u_D(k-2) = 4.4444 E_0 (e(k) - e(k-1))$$

$$\Rightarrow u_D(k) = (1+z_0)u_D(k-1) - z_0 u_D(k-2) + 4.4444E_0 e(k) - 4.4444E_0 e(k-1);$$
$$e(k) = 1 \quad \forall k \in [1, +\infty[$$

3 - According to the equation $u_D(k)$, the pole z_0 has an influence on the output controller.

Application 3 - PI controller

Consider a linear-time invariant system given by the following expression:

$$H(s) = \frac{1}{1+s}$$

followed by a zero-order-hold with sampling period $T_s = 0.6s$.

1- to synthesis a proportional integral control with:

- an overshoot less than 10%
- a settling time =7s
- a reference $y_r = 100$
- a damping ratio $\xi = 0.7$
- a natural frequency $w_n = 0.82$.

2- Simulate the discrete-time controller with the discretized plant to see whether the specifications are fulfilled in the discrete-time setting.
3- Simulate the discrete controller with the continuous system followed by a zero-order hold.

Answer

The transfer function of the discrete-time system is

$$H(z) = \left(1 - z^{-1}\right) Z\left[\frac{H(s)}{s}\right] = \left(1 - z^{-1}\right) Z\left[\frac{1}{s(1+s)}\right] = \frac{0.4512}{z - 0.5488}$$

the conventional version of PI controller is

$$C_{PI}(z) = \frac{(K_P + K_I)z - K_P}{z-1}$$

the transfer function in the open loop is

$$H_{OL}(z) = \frac{\left[(K_P + K_I)z - K_P\right]}{z-1}\frac{0.4512}{z-0.5488}$$

and the transfer function in the closed loop is

$$H_{CL}(z) = \frac{\dfrac{(K_P + K_I)z - K_P}{z-1}\dfrac{0.4512}{z-0.5488}}{1+\dfrac{(K_P + K_I)z - K_P}{z-1}\dfrac{0.4512}{z-0.5488}}$$

$$= \frac{0.4512\left[(K_P + K_I)z - K_P\right]}{(z-1)(z-0.5488)+0.4512\left[(K_P + K_I)z - K_P\right]}$$

$$= \frac{0.4512\left[(K_P + K_I)z - K_P\right]}{z^2 + \left[0.4512(K_P + K_I) - 1.5488\right]z + 0.5488 - 0.4512K_P}$$

The desired continuous closed-loop transfer function

$$H_{desired}(s) = \frac{w_n^2}{s^2 + 2\xi w_n s + w_n^2} = \frac{0.67}{s^2 + 1.15s + 0.67}$$

with a damping ratio $\xi = 0.7$ and a natural frequency $w_n = 0.82$.
The desired discrete closed-loop transfer function is

$$H_{desired}(z) = \frac{(1-z_p)(1-\overline{z}_p)}{(z-z_p)(z-\overline{z}_p)} = \frac{1-(\overline{z}_p + z_p) + z_p\overline{z}_p}{z^2 - z(z_p + \overline{z}_p) + z_p\overline{z}_p}$$

where $z_p = e^{sT_s} = e^{T_s\left(-\xi w_n \pm j w_n\sqrt{1-\xi^2}\right)} = e^{-T_s\xi w_n} e^{\pm jT_s w_n\sqrt{1-\xi^2}}$ so we find

$$H_{desired}(z) = \frac{0.1716}{z^2 - 1.3306z + 0.5022}$$

We determine the parameters of the proportional integral controller by comparing the denominator of the actual closed-loop transfer function

$$\begin{cases} 0.5488 - 0.4512K_P = 0.5022 \\ \left[0.4512(K_P + K_I) - 1.5488\right] = -1.3306 \end{cases}$$

so

$$\begin{cases} K_P = 0.1033 \\ K_I = 0.4348 \end{cases}$$

so the parameters of the PI controller are:

$$C_{PI}(z) = \frac{0.5381z - 0.1033}{z-1}$$

2- We simulate the discrete-time controller with the discretized plant to see whether the specifications are fulfilled in the discrete-time setting:

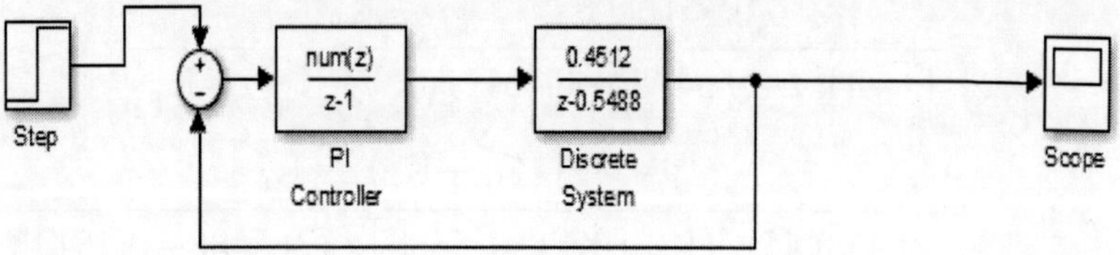

Fig. 5.38. The Simulink block of the system with PI controller

The curve of the output is given by the following figure

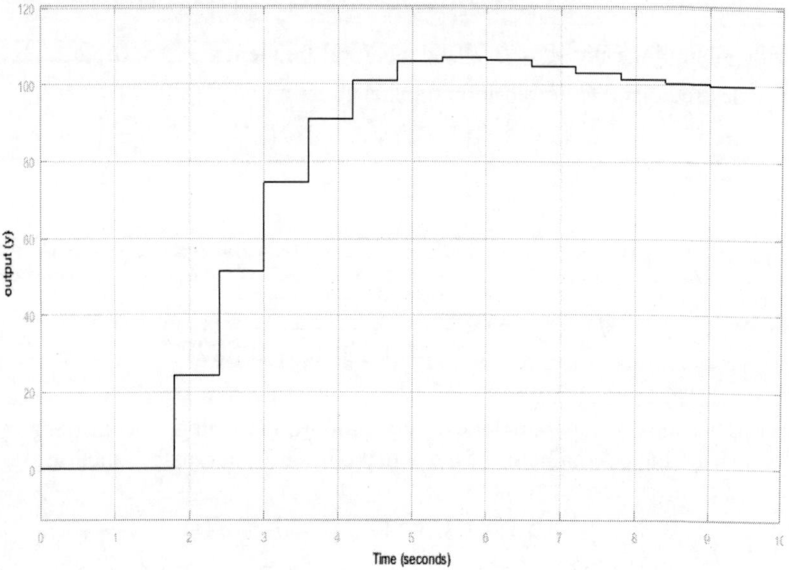

Fig. 5.39. The curve of the output

We remark that the overshoot seems slightly larger than the design specification. But the settling time meets the specification.

3- We simulate the discrete proportional integral controller with the continuous system followed by a zero-order hold.

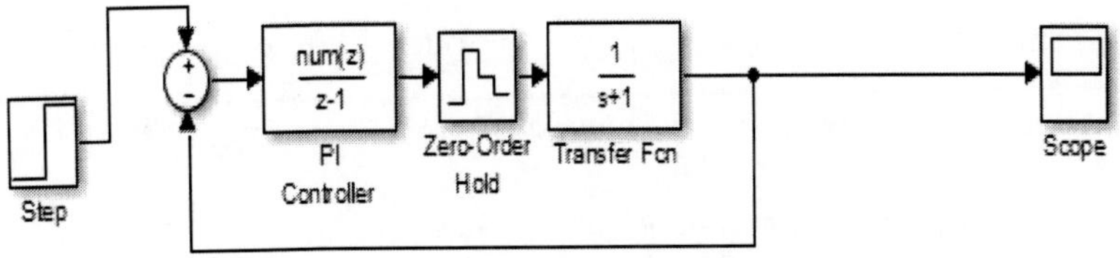

Fig. 5.40. The Simulink block of the discrete-time system with PI controller

The curve of the output is given by the following figure

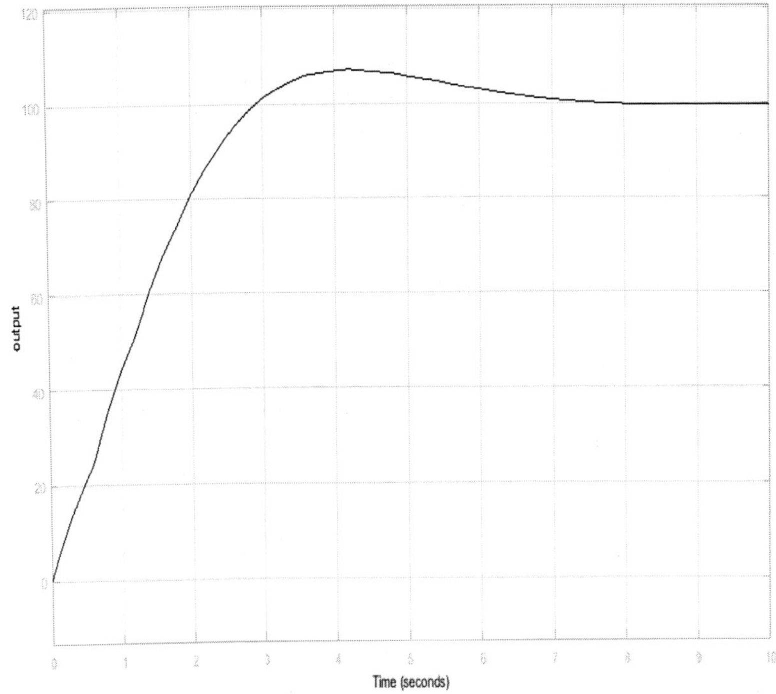

Fig. 5.41. The curve of the output

We remark that the overshoot and the settling time are above the same as those obtained with the discretized system. All design specifications are met with a sampling period Ts = 0.6 seconds.

Application 4 - PID controller

We want to apply a discrete-time controller PID to a second order continuous system with transfer function $H(s)=\dfrac{1}{s(s+1)}$ followed by a zero-order hold with sampling period $T_s=1s$.

1- Find the transfer function $H(z)$ of the system as the following expression

$$H(z)=\frac{b_1z^{-1}+b_2z^{-2}}{a_0+a_1z^{-1}+a_2z^{-2}}.$$

2- Find the closed loop transfer function.

3- Compute the parameters of the discrete-time PID controller using the following characteristic:

- damping ration $\xi=0.5$.
- natural frequency $w_n=2\pi\times 0.1$.

Answer

1 - The transfer function of the discrete-time system is

$$H(z)=(1-z^{-1})Z\left[\frac{H(s)}{s}\right]=(1-z^{-1})Z\left[\frac{1}{s^2(s+1)}\right]=\frac{0.368z+0.264}{z^2-1.268z+0.368}$$

$$H(z)=\frac{0.368z^{-1}+0.264z^{-2}}{1-1.268z^{-1}+0.368z^{-2}}=\frac{b_1z^{-1}+b_2z^{-2}}{a_0+a_1z^{-1}+a_2z^{-2}}$$

with

$$\begin{cases} b_1 = 0.368 \\ b_2 = 0.264 \\ a_0 = 1 \\ a_1 = -1.268 \\ a_2 = 0.368 \end{cases}$$

2 - In the closed-loop the performance specification of the system are:

$$\xi = 0.5; \quad w_n = 0.628 \quad \Rightarrow \quad p_{1,2} = -0.314 \pm 0.544\,j$$

$$z = e^{pT_s}; \quad \mathrm{T_s} = 1s \quad \Rightarrow \quad \mathrm{z}_{1,2} = 0.626 \pm 0.379\,j$$

The denominator of the system in the closed loop:

$$D_{CL}(z^{-1}) = (1 - z_1 z^{-1})(1 - z_2 z^{-1}) = 1 - 1.256 z^{-1} + 0.536 z^{-2}$$

$$\rightarrow \quad p_1 = -1.256, \qquad p_2 = 0.536$$

3 - Computing the parameters of discrete-time PID

$$D_{CL}(z^{-1}) = D(z^{-1})S(z^{-1}) + N(z^{-1})R(z^{-1})$$

$$D_{CL}(z^{-1}) = 1 + (a_1 - p - 1 + b_1 r_0) z^{-1} + (p - a_1 p - a_1 + a_2 + b_1 r_1 + b_2 r_0) z^{-2} +$$
$$(a_1 p - a_2 p - a_2 + b_1 r_2 + b_2 r_1) z^{-3} + (a_2 p + b_2 r_2) z^{-4} = 1 + p_1 z^{-1} + p_2 z^{-2}$$

$$D_{CL}(z^{-1}) = 1 + p_1 z^{-1} + p_2 z^{-2}$$

By identification we find

$$\begin{cases} p_1 = a_1 - p - 1 + b_1 r_0 \\ p_2 = p - a_1 p - a_1 + a_2 + b_1 r_1 + b_2 r_0 \\ 0 = a_1 p - a_2 p - a_2 + b_1 r_2 + b_2 r_1 \\ 0 = a_2 p + b_2 r_2 \end{cases}$$

$$\begin{cases} -1.256 = -1.368 - p - 1 + 0.368 r_0 \\ 0.536 = p + 1.368 p + 1.368 + 0.368 + 0.368 r_1 + 0.264 r_0 \\ 0 = -1.368 p - 0.368 p - 0.368 + 0.368 r_2 + 0.264 r_1 \\ 0 = 0.368 p + 0.264 r_2 \end{cases}$$

$$\begin{pmatrix} 0.368 & 0 & 0 & -1 \\ 0.264 & 0.368 & 0 & 2.368 \\ 0 & 0.264 & 0.368 & -1.736 \\ 0 & 0 & 0.264 & 0.368 \end{pmatrix} \begin{pmatrix} r_0 \\ r_1 \\ r_2 \\ p \end{pmatrix} = \begin{pmatrix} 1.112 \\ -1.2 \\ 0.368 \\ 0 \end{pmatrix}$$

$$\begin{pmatrix} r_0 \\ r_1 \\ r_2 \\ p \end{pmatrix} = \begin{pmatrix} 0.368 & 0 & 0 & -1 \\ 0.264 & 0.368 & 0 & 2.368 \\ 0 & 0.264 & 0.368 & -1.736 \\ 0 & 0 & 0.264 & 0.368 \end{pmatrix}^{-1} \begin{pmatrix} 1.112 \\ -1.2 \\ 0.368 \\ 0 \end{pmatrix} = \begin{pmatrix} 1.9249 \\ -2.0445 \\ 0.5626 \\ -0.4036 \end{pmatrix}$$

Finally, we find

$$\begin{cases} K_P = \dfrac{-r_0 p - r_1 - (2-p)r_2}{(1-s_1)^2} \\ T_I = T_s \dfrac{K_P(1-p)}{r_0 + r_1 + r_2} \\ T_D = T_s \dfrac{p_1^2 r_0 + p r_1 + r_2}{K_P(1-p)^3} \\ \dfrac{T_D}{N} = \dfrac{pT_s}{1-p} \end{cases}$$

$$\begin{cases} K_P = \dfrac{1.9249 * 0.4036 + 2.0445 - (2 + 0.4036) * 0.5626}{(1+0.4036)^2} \\ T_I = T_s \dfrac{K_P(1+0.4036)}{1.9249 - 2.0445 + 0.5626} \\ T_D = T_s \dfrac{(0.4036))^2 * 1.9249 + 0.4036 * 2.0445 + 0.5626}{K_P(1+0.4036)^3} \\ \dfrac{T_D}{N} = \dfrac{-0.4036 T_s}{1+0.4036} \end{cases}$$

$$\begin{cases} K_P = -0.2921 \\ T_I = -0.9255 \\ T_D = -2.1063 \\ \dfrac{T_D}{N} = -0.2875 \Rightarrow \text{N}=7.3263 \end{cases}$$

5.7 Conclusions

In this chapter, we aimed to explain how can we successfully use P, PI, PD and PID controllers in many applications. Indeed, we tried to focus on almost all aspects of PID control in discrete-time linear invariant system because PID controllers and their variations are commonly used in the industry. Control engineers usually prefer PI controllers to control first order system and, PID control is vastly used to control two or higher order systems. In almost all cases fast transient response and zero steady state error is desired for a closed loop system. Usually, these two specifications conflict with each other which makes the design harder. The reason why PID is preferred is that it provides both of these features at the same time.

5.8 References

Ostalczyk, P. 2018. Variable, fractional-order RST/PID controller transient characteristics calculation. IEEE, 23rd International Conference on Methods & Models in Automation & Robotics: 93-98.

Ding, Z. 2013. Nonlinear and adaptive control systems: The Institution of Engineering and Technology. Institution of Engineering and Technology.

Aguirre, L.A. 1992. PID tuning based on model matching, IEEE electronic letter, Vol. 28, No. 25: 2269-2271.

Gaing, Z.L. 2004. A particle swarm optimization approach for optimum design of PID controller in AVR system, IEEE transaction on energy conversion, Vol. 19, No. 2: 384-391.

Hauksdottir, A.S., Herjolfsson, G., Sigurdsson, S.T. 2007. Optimizing zero tracking and disturbance rejecting controllers - The generalized PID controller. The 2007 American Control Conference. New York City, USA, July 11-13: 5790–5795.

Shankar, S. and Marc, B. 1989. Adaptive control: stability, convergence, and robustness: Prentice Hall information and system sciences series.

Herjolfsson, G. and Hauksdottir, A.S. 2004. Direct computation of optimal discrete time PID controllers. The 2004 American Control Conference. Boston, Massachusetts, June 30 - July 2: 46-51.

Astrom, K. J. & Hagglund, T. 1995. PID controllers: theory, design and tuning, International Society for Measurement and Control, ISBN 1-55617-516-7, Durham, North Carolina, USA.

Aström, K.J. and Haglund, T. 1995. PID Controllers: Theory, Design and Tuning. Instrument Society of America.

Gang, F. and Rogelio, L. 1999. Adaptive control systems. Butterworth-Heinemann Linacre House, Jordan Hill, Oxford OX2 8DP.

O'Dwyer, A. 2006. Handbook of PI and PID controller tuning rules, Imperial College Press.

Boyd, S. and Barratt, C. 1991. Linear controller design, limits of performance. Prentice Hall.

Adusumilli, S., Rivera, D., Dash, S. and Tsakalis, K. 1998. Integrated MIMO identification and robust PID controller design through loop shaping. Proceedings of the 1998 American Control Conference.

Derek, A. 2014. An introduction to nonlinearity in control systems: bookboon.com ISBN 978-87-7681-790-9.

Grassi, E., Tsakalis, K., Das, S., Gaikwad, S. and Stein, G. 2000. Adaptive/self-tuning PID control by frequency loop-shaping. Proc. Of 39th IEEE CDC.

Grassi, E. and Tsakalis, K. 1996. PID controller tuning by frequency loop-shaping. Proc. 35th Conference on Decision and Control.

Grassi, E., Tsakalis, K., Dash, S., Gaikwad, S., MacArthur, W. and Stein, G. 2001. Integrated system identification and PID controller tuning by frequency loop-shaping. IEEE Transactions on Control Systems Technology, Vol. 9, No. 2: 285-294.

Bruce, P. 2017. Essential process control for chemical engineers: bookboon.com ISBN 978-87-403-1655-1.

Voda, A. and Landau, I. 1995. A method for auto-calibration of PID controllers. Automatica, Vol. 31, Issue 1.

Nudelman, G. and Kulesky, R. 2001. New approach to PID-control optimization for power station loops. Proceedings Of European Control Conf.

Kristiansson, B. and Lennartson, B. 2002. Robust and optimal tuning of PI and PID controllers. Control Theory and Applications, IEE Proceedings, Vol. 149, No. 1: 17-25.

Malan, S., Milanese, M. and Taragna, M. 1994. Robust tuning for PID controllers with multiple performance specifications. Proceedings of the 33rd IEEE Conference on Decision and Control.

Derek, A. 2015. Control engineering problems with solutions: bookboon.com ISBN 978-87-403-0374-2.

Astrom, K.J. and Hagglund, T. 1998. Automatic tuning of PID controllers. Instrument Society of America.

Grassi, E. and Tsakalis, K. 2000. PID controller tuning by frequency loop-shaping: application to diffusion furnace temperature control. IEEE Transactions on Control Systems Technology, Vol. 8, No. 5: 842-847.

Arild, W. 2013. Discrete dynamical systems with an introduction to discrete optimization problems: bookboon.com ISBN 978-87-403-0327-8.

Karpiš, O., Miˇcek, J., Olešnaníková, V. 2015. Using of compressed sensing in energy sensitive WSN applications, FedCSIS 2015, September 13-16, Lodz, Poland ISBN 978-83-60810-65-1, pp. 1233-1238, http://dx.doi.org/10.15439/2015F167.

Skovranek, T., Despotovic, V. 2009. Identification of systems of arbitrary real order: a new method based on systems of fractional order differential equations and orthogonal distance fitting, ASME 2009 Inter. Design En. Tech. Conf. and Computers and Information in Engineering Conference: pp. 1063-1068, 2009.

McKelvey, T. 2004. Least squares and instrumental variable methods, control systems, robotics, and automation, in EOLSS, Developed under the auspices of the UNESCO 2004.

Guo, W. 2014. Modelling and simulation of a capacitive micro-accelerometer system, Proceedings of the 33rd Chinese Control Conference July, Nanjing, China, 2014